Springer Tracts on Transportation and Traffic

Volume 24

About this Series

The book series "Springer Tracts on Transportation and Traffic" (STTT) publishes current and historical insights and new developments in the fields of Transportation and Traffic research. The intent is to cover all the technical contents, applications, and multidisciplinary aspects of Transportation and Traffic, as well as the methodologies behind them. The objective of the book series is to publish monographs, handbooks, selected contributions from specialized conferences and workshops, and textbooks, rapidly and informally but with a high quality. The STTT book series is intended to cover both the state-of-the-art and recent developments, hence leading to deeper insight and understanding in Transportation and Traffic Engineering. The series provides valuable references for researchers, engineering practitioners, graduate students and communicates new findings to a large interdisciplinary audience.

Indexed by SCOPUS, WTI Frankfurt eG, zbMATH, SCImago.

Rishit Agnihotri

Electrified Road Systems (ERS) for Emerging Asia

Springer

Rishit Agnihotri
Department of Computer
and Communication Engineering
Manipal University Jaipur
Jaipur, Rajasthan, India

ISSN 2194-8119 ISSN 2194-8127 (electronic)
Springer Tracts on Transportation and Traffic
ISBN 978-3-032-16926-6 ISBN 978-3-032-16927-3 (eBook)
https://doi.org/10.1007/978-3-032-16927-3

This Springer imprint is published by the registered company Springer Nature Switzerland AG
The registered company address is: Gewerbestrasse 11, 6330 Cham, Switzerland

Competing Interests The author has no competing interests to declare that are relevant to the content of this manuscript.

Contents

1 Introduction to Electric Road Systems (ERS) 1

1.1 Background and Motivation 1

1.1.1 Global Imperatives: Decarbonization and the Mobility Transition 1

1.1.2 Limitations of Battery-Only Electric Vehicle (BEV) Systems 2

1.1.3 Introduction to Electric Road Systems (ERS) as a Paradigm Shift 3

1.2 Definition and Scope of ERS 4

1.2.1 Conceptual Definition and Core Functionality of ERS 4

1.2.2 Classification of ERS Technologies 4

1.2.3 Areas of Application and Strategic Deployment Scenarios 5

1.3 Technological Foundations of ERS 6

1.3.1 Conductive Electric Road Systems (C-ERS) 6

1.3.2 Inductive Electric Road Systems (I-ERS or DWPT) 7

1.3.3 Capacitive Electric Road Systems (CAP-ERS or CPT) 8

1.3.4 Comparative Analysis of ERS Technologies 9

1.4 Benefits and Potential Impacts 10

1.4.1 Environmental and Energy Impact 10

1.4.2 Economic Benefits and Life-Cycle Cost (LCC) Reduction 11

1.4.3 Operational and Societal Impact 12

1.5 Challenges and Barriers 13

1.5.1 Technical and Engineering Challenges 13

1.5.2 Economic, Regulatory, and Policy Challenges 13

1.5.3 Social Acceptance and Adoption Barriers 15

1.6 Global and Regional ERS Initiatives 15
1.6.1 Major ERS Deployments in Europe and North America 15
1.6.2 Focus on Asia: Korea's Commercial Trials and Policy Initiatives 16
1.6.3 Policy and Research Acceleration in India and the Broader Asian Context 16
1.7 Research Gaps and Motivation for Study 17
1.7.1 Critical Technical Gaps 17
1.7.2 Economic Gaps 18
1.7.3 Policy and Governance Gaps 18
1.7.4 Justification for the Monograph 19
1.8 Chapter Summary and Monograph Transition 19
References 20

2 Global Technological Landscape of ERS 23
2.1 Introduction 23
2.1.1 Defining the ERS Technology Portfolio 24
2.1.2 Historical Context and Evolution 24
2.2 ERS Technologies—Technical Overview 25
2.2.1 Overhead Conductive Systems 26
2.2.2 Ground-Level Conductive Systems 27
2.2.3 Wireless Dynamic Charging (Inductive) 29
2.3 Global Pilot Projects 32
2.3.1 Sweden—Smart Road Gotland (Electreon) 32
2.3.2 Germany—Electreon Trials 33
2.3.3 United States—California eHighway (Siemens) 34
2.3.4 South Korea—Gyeonggi Province Test Bed (KAIST OLEV) 35
2.4 Comparative Analysis of ERS Technologies 36
2.4.1 Energy Efficiency 38
2.4.2 Installation and Maintenance Costs 38
2.4.3 Scalability and Application Suitability 39
2.4.4 Safety and Environmental Impact 40
2.5 Funding Models 40
2.5.1 Public–Private Partnerships (PPPs) 41
2.5.2 Government Grants and Public Funding 42
2.5.3 Private Investment and Commercial Models 42
2.5.4 Cost-per-Kilometre Analysis 43
2.6 Challenges and Considerations 44
2.6.1 Technical Challenges 44
2.6.2 Economic Challenges 45
2.6.3 Social and Regulatory Challenges 46

2.7 Future Outlook and Research Gaps 47
2.7.1 Emerging Technologies and Efficiency Improvements 47
2.7.2 Integration with Smart Grids and Renewable Energy 47
2.7.3 Scalability Pathways and Policy Frameworks 48
2.7.4 Key Research Gaps 49
2.8 Conclusion 50
References 51

3 Dynamic Wireless Charging: Principles and Applications 53
3.1 Introduction 53
3.2 Fundamental Principles of DWPT 54
3.2.1 Electromagnetic Induction and Magnetic Resonance Coupling 54
3.2.2 Comparative Analysis: Dynamic Versus Static Wireless Charging 55
3.3 System Elements and Architecture 56
3.3.1 Overall System Architecture 56
3.3.2 Roadside Infrastructure: The Transmitter System 58
3.3.3 Vehicle-Side Components: The Receiver System 59
3.3.4 Compensation Topologies 60
3.4 Recent Developments and Innovations (2020–2025) 62
3.4.1 Advanced Coil Designs for Scalability and Efficiency 62
3.4.2 Interoperability and Standardisation 63
3.4.3 Advanced Materials and AI-Based Optimisation 64
3.5 Applications and Deployment Scenarios 65
3.5.1 Passenger EVs: Range Extension and Battery Downsizing 65
3.5.2 Urban Public Transit: Continuous Charging Corridors 66
3.5.3 Long-Haul Freight and Logistics: Reducing Downtime 67
3.5.4 Other Applications: Port Automation and Autonomous Vehicles 67
3.5.5 Global Pilot Projects and Performance Metrics 68
3.6 Challenges and Mitigation Strategies 68
3.6.1 Technical Challenges 68
3.6.2 Economic Viability 70
3.6.3 Safety and Regulatory Compliance 71
3.7 Future Outlook 72
3.7.1 Integration with Smart Grids and Renewable Energy 73

3.7.2 Synergy with Autonomous Vehicles and V2X Communication 73
3.7.3 Roadmap for Global Scaling Towards 2040 74
3.8 Conclusion 74
References 76

4 Grid Integration and Energy Management in Electric Road Systems (ERS) 79
4.1 Introduction 79
4.1.1 Defining Grid Integration in the ERS Context 79
4.1.2 The Importance of Real-Time Power Exchange 80
4.1.3 The Role of Smart Grids and Distributed Energy Resources (DER) 81
4.1.4 Chapter Roadmap 81
4.2 Power Flow and Grid Architecture in ERS 82
4.2.1 Grid Topology for ERS 82
4.2.2 Power Conversion, Voltage Regulation, and Power Flow Control 85
4.2.3 Bidirectional Power Flow (V2G) in ERS 85
4.3 Load Balancing and Peak Demand Management 86
4.3.1 Dynamic Load Balancing (DLB) 87
4.3.2 Demand Response (DR) and Time-of-Use (TOU) Scheduling 87
4.4 Integration of Renewable Energy Sources 90
4.4.1 Architectures for Renewable Integration and the Role of Microgrids 90
4.4.2 Energy Storage Solutions for Mitigating Intermittency 91
4.4.3 Case Study: Applying Solar Microgrid Principles to ERS 91
4.5 Energy Pricing and Economic Management 93
4.5.1 Concepts in Dynamic Energy Pricing 93
4.5.2 Benefits and Models for ERS Economic Management 94
4.6 AI-Based Energy Routing and Edge-Grid Coordination 95
4.6.1 The Role of AI/ML in Predictive Load Management 96
4.6.2 Advanced Algorithms for Energy Routing and Control 96
4.6.3 Edge Computing and the Internet of Energy (IoE) 97
4.7 Challenges and Reliability Issues 98
4.7.1 Technical Challenges 99
4.7.2 Operational and Economic Challenges 99
4.7.3 Cyber-Physical Vulnerabilities 100

4.8 Future Trends and Research Directions ... 101
4.8.1 Grid Decentralisation and Roadway Microgrids ... 101
4.8.2 Blockchain-Based Peer-to-Peer Energy Exchange ... 102
4.8.3 Fully Integrated Vehicle-to-Everything (V2X) Networks ... 102
4.8.4 Autonomous Grid Balancing with AI and Digital Twins ... 103
4.9 Conclusion ... 104
References ... 104

5 Government Policy Frameworks and Roadmaps for Electric Road Systems (ERS) ... 107
5.1 Introduction ... 107
5.2 Policy Landscape for ERS and E-Mobility ... 108
5.3 Public–Private Partnership (PPP) Models ... 109
5.4 Regulatory Frameworks and Standards ... 112
5.4.1 Safety and Electromagnetic Compliance (EMC) ... 113
5.4.2 Interoperability and Standardisation ... 115
5.4.3 Road Access, Urban Planning, and Public Space ... 116
5.5 Case Studies: National ERS Roadmaps ... 116
5.5.1 Germany ... 117
5.5.2 Sweden ... 118
5.5.3 United States ... 119
5.5.4 South Korea ... 120
5.6 Policy Instruments and Economic Incentives ... 120
5.7 Institutional Coordination and Governance ... 122
5.8 Policy Framework Proposal for India and Asia ... 125
5.8.1 Pillar 1: Legislation and Standards Alignment ... 125
5.8.2 Pillar 2: Financing and PPP Enablement ... 126
5.8.3 Pillar 3: Pilot Corridors and Regional Prioritisation ... 127
5.8.4 Pillar 4: Capacity Building and Skill Development ... 128
5.8.5 Pillar 5: Digital and Grid Integration ... 128
5.9 Future Outlook ... 128
5.10 Conclusion ... 130
References ... 131

6 ERS for Emerging Asia: Infrastructure Readiness and Challenges ... 135
6.1 Introduction ... 135
6.2 Regional Overview of Road Infrastructure ... 136
6.3 Civil Engineering and Construction Standards ... 138
6.4 Electrical and Grid Infrastructure Readiness ... 139
6.5 Freight Logistics and Traffic Patterns ... 142
6.6 Road Quality and Safety Indicators ... 143

6.7 ERS Readiness Index (ERI) and Comparative Assessment 144
6.8 Challenges and Bottlenecks 146
6.9 Strategic Interventions and Adoption Pathways 148
6.10 Conclusion 151
References 152

7 Cost–Benefit Analysis and Financing Models for Electric Road Systems (ERS) 157
7.1 Introduction: The Economic Rationale for Dynamic Charging 157
7.1.1 A Life-Cycle Cost Framework for ERS 158
7.1.2 Deconstructing Capital Expenditures (CapEx) 158
7.1.3 Analysing Operational Expenditures (OpEx) 158
7.2 Comparative Cost Analysis of ERS Technologies 159
7.3 Socio-Economic Cost–Benefit Analysis (CBA) 160
7.3.1 Methodological Foundations 161
7.3.2 Direct and Indirect Economic Benefits 161
7.4 Monetising Externalities 162
7.5 Comparative CBA: ERS vs. Alternative Decarbonization Pathways 162
7.6 Financing and Investment Models for ERS Deployment 164
7.6.1 The Role of Public Sector Funding 164
7.6.2 Public–Private Partnership (PPP) Structures 165
7.6.3 Attracting Private Capital and Green Finance 166
7.7 Strategic Recommendations for Financial and Policy Frameworks 167
References 168

8 Integration of Electric Road Systems with Public Transport and Freight Networks 173
8.1 Introduction 173
8.2 ERS Integration in Public Transport 174
8.2.1 Operational Advantages: Continuous Charging and Fleet Optimisation 174
8.2.2 Infrastructure Considerations and System Design 175
8.2.3 Case Example: Evolution Road in Lund, Sweden 175
8.3 Integration with Urban and Intercity Freight Networks 177
8.3.1 The Hybrid Operational Model: A Transitional Framework 177
8.4 Technical Requirements for Integration 179
8.4.1 Road Infrastructure Adaptation 180
8.4.2 Fleet Management and Telematics 181
8.5 Operational and Economic Benefits 181
8.5.1 Reduction in Fleet Battery Size and Cost 181
8.5.2 Reduced Charging Downtime and Improved Asset Utilisation 182

8.5.3 Lower Energy and Maintenance Expenditure 183
8.5.4 Quantified Emission Reductions 183
8.6 Urban Planning and Infrastructure Integration 183
8.6.1 Integration with Smart City Systems 184
8.6.2 ERS in City Master Plans and Transit-Oriented Development (TOD) 185
8.6.3 Collaboration Between City Transport Authorities and Utilities 185
8.7 Case Studies and Global Best Practices 186
8.8 Integration Framework for Emerging Asia 187
8.8.1 A Phased Regional Integration Model 189
8.8.2 Institutional Roles and Collaboration 190
8.9 Challenges and Mitigation Strategies 190
8.9.1 Future Outlook 194
8.9.2 Integration with Digital Twins and AI-Based Optimisation 194
8.9.3 ERS and Vehicle-To-Grid (V2G) Logistics Nodes 194
8.9.4 Policy Synergy with Smart City Integration 195
8.9.5 Conclusion 195
References 196

9 Urban and Smart City Integration of Electric Road Systems (ERS) 199
9.1 Introduction 199
9.2 ERS in the Smart City Ecosystem 201
9.3 IoT-Based Traffic and Energy Management 202
9.3.1 The Cyber-Physical System of Urban ERS 203
9.3.2 Intelligent Traffic Management Applications 204
9.3.3 Smart Grid Integration and Energy Synchronisation 204
9.4 Data Architecture and Communication Framework 205
9.4.1 Structure of the Integrated Data Network 206
9.4.2 Key Communication Standards and Protocols 207
9.4.3 Role of Advanced Digital Technologies 208
9.5 Integration with City-Level Mobility Systems 209
9.5.1 Electrifying Public Transport Corridors 209
9.5.2 Optimising Municipal and Commercial Fleets 210
9.5.3 Supporting Micro-mobility and Intermodal Connectivity 210
9.6 Urban Planning and Infrastructure Readiness 212
9.6.1 Incorporating ERS in Urban Master Plans 212
9.6.2 Institutional Coordination and Physical Modifications 212
9.6.3 Link with Sustainable Urban Development Principles 213

9.7 Policy and Governance Mechanisms 214
9.7.1 Local Policy Instruments for ERS Adoption 214
9.7.2 Institutional Models for Implementation and Operation 215
9.7.3 Governance Challenges 216
9.8 Case Studies 216
9.8.1 Stockholm, Sweden (eRoadArlanda) 217
9.8.2 Busan/Gumi, South Korea (OLEV) 218
9.8.3 Berlin/Frankfurt, Germany (eHighway) 218
9.8.4 Potential in Pune or Bengaluru, India 219
9.9 Future Smart City Framework for Asia 219
9.9.1 Pillars of the Conceptual Model 220
9.9.2 Linkages to National Digital Transformation Agendas 222
9.10 Challenges and Opportunities 225
9.10.1 Technical Challenges 225
9.10.2 Economic Challenges 225
9.11 Conclusion 227
References 228

10 Conclusions and Future Research Directions: Charting the Path to a Systemic Mobility Transition 231
10.1 Overview of Key Findings: A System-Of-Systems Perspective 231
10.2 Technological Evolution and Maturity: From Pilot to Practice 233
10.3 Economic and Policy Insights: The Business Case for Electrification 235
10.4 Environmental and Social Impacts: Aligning with Global Sustainability Goals 237
10.5 Global Strategic Implications: ERS as an Enabler of the Net-Zero Transition 239
10.6 Future Research Directions: An Interdisciplinary Agenda for 2025–2045 240
10.7 Roadmap for the Future: A Phased Global Rollout 245
10.8 Final Remarks: A Systemic Innovation for a Sustainable Future 246
References 247

Chapter 1
Introduction to Electric Road Systems (ERS)

1.1 Background and Motivation

1.1.1 Global Imperatives: Decarbonization and the Mobility Transition

The twenty-first century has posed an unparalleled worldwide threat in the reduction of climate change, which requires a swift and extensive reorganisation of the world's energy consumption and delivery systems. The main focus of this transition is the overall decarbonisation of the transport sector, which is still a major source of global greenhouse gas (GHG) emissions. The transition to electric vehicles (EVs) has become one of the key priorities to meet the high climate targets outlined in global agreements, which offers a direct channel toward a tremendous decrease in GHG emissions as well as local air pollution. Scholarly research on the macro-level effects of EV implementation can prove that there is a net positive effect, as EV adoption decreases total carbon dioxide emissions [1].

This trend of electrification across the world depends essentially on the concomitant developments in the generation of energy. The effectiveness of electric mobility in cutting carbon emissions is inseparably connected with the grid energy constitution of the grid that provides the electricity. Studies that measure this synergy show that a one percent increase in the capacity to generate renewable energy can increase the decarbonisation advantages of EVs by 0.036%. This is an important relationship that will raise the planning and implementation of charging infrastructure to a new level of utility-scale energy management, i.e. Electric Road Systems (ERS). The conventional static charging, especially mass charging of heavy-duty vehicles (HDVs), produces concentrated peaks, which exacerbate stability issues with intermittent renewable sources, including the better-known duck curve effect. ERS, in its turn, has the innate ability to distribute the electrical load spatially on highways and over the day, which can be dynamically controlled. This architectural feature

R. Agnihotri, *Electrified Road Systems (ERS) for Emerging Asia*, Springer Tracts on Transportation and Traffic 24, https://doi.org/10.1007/978-3-032-16927-3_1

enables ERS to serve as not only a delivery system for vehicle power, but also as an enabling technology to the optimised, large-scale consumption of intermittent renewable energy, which otherwise would necessitate the financial cost of very expensive grid-scale battery storage systems necessary to stabilise the grid under mass static charge conditions.

Moreover, full sector decarbonisation cannot be achieved without the electrification of HDVs and long-haul freight transport, but this sector is disproportionately hit by energy issues. The nature of freight transport is highly energy-consuming, and the mass of the battery needed to cover long-distance routes would be too large and heavy. ERS proposes the concept of dynamic charging that is specifically designed to serve these long routes and high-demand routes [2, 3]. As such, ERS is no longer a side feature created to make passenger cars more convenient, but a prerequisite of deep decarbonisation of the logistics and freight sector, where battery-only systems have the most significant technological and economic constraints.

1.1.2 Limitations of Battery-Only Electric Vehicle (BEV) Systems

Although there are strong environmental advantages to switching to Battery-Electric Vehicles (BEVs), the general, large-scale implementation of the existing BEV technology faces several systemic and engineering bottlenecks that jeopardise the adoption rates, especially in terms of operational efficiency, long-range logistics, and electric grid stability [4]. The best-known obstacle is the psychological phenomenon of range anxiety, which forces consumers and commercial fleet operators to insist on EVs with larger and larger batteries to ensure working autonomy [4]. This quest for extended range would require the incorporation of heavy and energy-dense battery packs, which, in turn, would cause drastic economic and environmental trade-offs. The use of oversized batteries reduces the total life-cycle vehicle energy efficiency. An example is that more complex modelling shows a significant increase in vehicle energy consumption of between 13.4 and 16.9% as battery capacity is increased between 28 and 116 kWh, simply because the mass penalty of the larger pack exists [5]. This use of more energy contradicts the very purpose of electrification, maximum energy efficiency, and raises a considerable efficiency paradox. The mechanism selected to guarantee range (larger battery) also lowers the operational energy efficiency through unnecessary weight. ERS provides a way to resolve this dilemma by separating the driving range and the storage capacity on board, so that vehicles can be right-sized to meet their daily buffer needs, thus doing so with maximum efficiency by reducing unnecessary mass.

Besides operational fines, huge battery packs pose a major bottleneck on the charging infrastructure and cause acute grid dependency problems. Heavy-duty vehicles, specifically, demand huge energy sources, which can be translated into significantly long periods to attain a full charge [6]. The deployment of the strong, high-power stationary charging infrastructure required to satisfy this demand causes sustained high-power consumption peaks that place unnecessary burden on the local distribution networks, potentially overloading local transformers and utility feeders [6]. The implementation of the ERS along significant corridors reduces this risk because instead of the localised distribution systems (DSOs) bearing the cost of bulk energy consumption, the transmission backbone will bear the same cost. Power management, stability, and voltage regulation in such high-voltage systems are more easily scalable and controllable, meaning that a paradigm change in utility planning methodology is required.

Moreover, the original high price of big batteries is a major contributor to the purchase price of the vehicle. In addition to the initial capital cost, the life-cycle cost (LCC) of the battery alone presents complicated economic factors. Studies on vehicle-to-grid applications and more general EV degradation models have found that the most common degradation cost is battery degradation caused by calendar ageing, the inability of the irreversible capacity loss over time, irrespective of use. A single extensive study revealed that calendar ageing explained 67% of the degradation expenses after 10 years, especially in low-mileage conditions [7]. This fact is a strong indication that large battery capacity gained with the sole purpose of assuring range is often not fully used, thus resulting in poor capital efficiency. ERS helps avoid this problem by lowering the required battery capacity and reducing the capital allocated to the under-utilised assets.

1.1.3 Introduction to Electric Road Systems (ERS) as a Paradigm Shift

Battery mass-related constraints, infrastructural strain-related constraints and the efficiency paradox all necessitate a re-evaluation of the energy delivery paradigm of electric mobility. Electric Road Systems (ERS) go further to promote a revolutionary solution, which enables the dynamic charging of a vehicle and constant renewal of energy when the vehicle is in motion, and this approach addresses the fundamental constraints of battery-centric BEVs [2].

ERS reinvests the road network, per se, as an energy infrastructure. An equivalent solution to the range-anxiety obstacle is to offer ubiquitous, continuous charging along major transportation routes, by which ERS can essentially remove the range-anxiety barrier and permit a significant reduction in corresponding increase in the required onboard battery capacity [3, 4]. The strategic reduction has many-fold advantages that are translated directly into lower vehicle prices, fewer material requirements, and increased energy efficiency due to less mass. Dynamic charging, therefore, represents an essential technological change, allowing the growth of electric mobility not only of personal cars but also of the sector that is extremely energy-demanding, heavy freight logistics.

1.2 Definition and Scope of ERS

1.2.1 Conceptual Definition and Core Functionality of ERS

Electric Road Systems (ERS) are fully integrated and systemised infrastructure that is capable of providing electric vehicles (EVs) with electrical energy on the move, thus enabling what is commonly referred to as Dynamic Wireless Power Transfer (DWPT) or dynamic charging. The fundamental idea of ERS is to change the basic nature of the working relationship between the EV and its source of power. The onboard battery of a vehicle in an ERS system will no longer be the main component that allows the car to cover the full driving range; it will only act as an energy store to maintain the vehicle's operation when it is not on the electrified section or in a low-speed manoeuvre [8].

This feature enables ERS to overcome the range and weight constraints that are typical of BEV systems. Nevertheless, the use of ERS requires more than a charging mechanism. ERS can be described as a sophisticated system-of-systems [2], which requires very high standards of integration in a variety of fields, such as advanced civil engineering to modify the roads, powerful power electronics to convert high-frequency energy and complex communication protocols (Vehicle-to-Infrastructure and standardised vehicle management interfaces). The effective functioning of ERS depends on instantaneous and smooth communication between the vehicle and the road infrastructure to start, support, measure and conclude power transmission reliably at high velocities.

1.2.2 Classification of ERS Technologies

The ERS technologies are categorised mainly by the physical principle used to convey electrical energy across the intervening gap (or air gap) between the fixed infrastructure embedded within or above the roadway and the collecting mechanism attached to the vehicle. The main category is divided into two major categories:

- Conductive ERS (C-ERS)
- Wireless ERS (W-ERS).

Conductive ERS (C-ERS) requires physical contact to transmit power; the vehicle has a current collector, a pantograph or a retractable rail collector (plough), to make physical contact with the roadside infrastructure, which is usually an overhead catenary wire or an embedded ground rail [3, 9]. This approach is very mature and builds on a hundred years of experience of electrification of railways and trolleys [9].

Wireless ERS (W-ERS) is a transfer of power over the air without any physical contact and is based on the electromagnetic fields generation and transfer. W-ERS is further classified by the main field used including Inductive Power Transfer (IPT), or Dynamic Wireless Power Transfer (DWPT), which uses resonant magnetic coupling and which transfers power between parallel conductive plates (one set embedded in the road, one set mounted on the vehicle) separated by a dielectric material (air and road material) [10].

Other than the transfer mechanism, ERS operational modes may be categorized by motion: Static Charging (standard plug-in or stationary WPT, e.g., during parking); Quasi-Dynamic Charging (charging during the vehicle stop or at slow speeds, common at bus stops or at a traffic light); and Fully Dynamic Charging (the primary functionality of ERS, where power transfer occurs continuously during high-speed operation on large highways or transit corridors).

1.2.3 Areas of Application and Strategic Deployment Scenarios

The high initial cost of ERS infrastructure will require a planned rollout with a view towards use cases that will have the most favourable operational and economic returns, usually in areas where the traditional BEV constraints are most significant.

Heavy-Duty Freight Corridors and Logistics

ERS is best applicable to long-haul trucking and heavy-duty vehicles (HDVs). In this type of vehicle, the amount of energy needed and the size of the battery needed to operate over a long period of time are usually prohibitive in terms, cost and charge time. Conductive ERS, in particular overhead catenary, is quite appropriate in such a situation due to its demonstrated efficiency (in most cases greater than 95) and ability to provide high, steady power required in propelling heavy loads [3, 9]. Planned electrification of key freight routes changes logistics by ensuring that they are operated on a 24/7 basis with zero emissions.

Cities' Public Transportation and Fleets

Metropolitan transit buses' routes and last-mile delivery fleets are marked by predictable and high-frequency travel on fixed and defined routes. ERS is very helpful in these applications since either conductive ground rails can be used or, more normally, Inductive ERS (I-ERS). I-ERS has the advantage of having seamless and buried infrastructure that does not visually intrude, as is the case with overhead lines in urban areas where the lines are dense. Moreover, the opportunities of quasi-, dynamic-, and charging bus stops or termini can be used by the vehicles of the public transport, which spend a short time there, providing the guaranteed maintenance of the batteries during the working day. The projects pioneered in South Korea have proved the application of I-ERS in public transportation.

Passenger Vehicle Integration

To the common consumer, the main addition to the variety of smaller-battery passenger cars is the ERS [11]. Long-range anxiety can be successfully overcome through the strategic use of major intercity routes and congested highway segments. With this dynamic charging feature, governments and infrastructure planners can make a bigger portion of the population adopt EVs, but without forcing consumers to buy battery packs that are disproportionately large, heavy, and expensive [2, 8, 11].

1.3 Technological Foundations of ERS

1.3.1 Conductive Electric Road Systems (C-ERS)

The Conductive Electric Road Systems (C-ERS) operate on conductive roads and tracks, which are poorly drained, such that they may flood after a fierce rainfall. C-ERS is currently the most technologically advanced design of dynamic power transfer, reaching the highest Technological Readiness Level (TRL) of variants of electric road system (ERS). This maturity is based on the fact that over a hundred years of experience in the successful electrification of the railway was gained [9].

Overhead Catenary Systems (OCS)

The most common arrangement of C-ERS, the OCS, is an overhead conductor infrastructure network suspended over the road surface. This system has a main contact wire, which has direct physical contact with the vehicle, a messenger wire located above it to distribute tension and offer support and hanger wires, which attach both ends of the system [9]. A roof-mounted current collector called a pantograph transfers the energy to the vehicle [3, 9]. The direct metallic connection provides very high system efficiencies, often exceeding 95–97%, since losses due to the magnetic or electric field generation and coupling are practically zero. The established trustworthiness of OCS also allows much quicker acceleration and deceleration of vehicles than traditional diesel propulsion, which improves the general efficiency of service [9].

Ground Rail Systems

Ground-rail C-ERS—these mount energy rails directly in the roadbed and are reached by a retractable, underslung collector or plough. Segmented rail activation is one of the key safety aspects of such systems and allows only the parts of the rail that are under the travelling object to be energised and thus eliminating the electrical risks to pedestrians, animals, and service workers. Ground-rail systems are faced with significant durability problems, although OCS is relatively resilient. The infrastructure should be strengthened against high mechanical loads caused by traffic loads, be made tolerant to debris, and have strong ingress resistance to water and chemicals like de-icing salts. Furthermore, snow and ice formation may easily disrupt the physical contact process, requiring complex heating or removal equipment. Vertical and lateral alignment is very important in the operational success of C-ERS, which can be overhead or rail-based. In the case of high-speed highway use, this requires very precise civil works and constant inspection of infrastructure integrity. Non-technical factors often form the major obstacle to generalised deployment of C-ERS into generalised public traffic, and this is the issue of societal and aesthetic acceptability, and as such is limited to specific dedicated freight transport corridors and heavy logistic centres, making it especially useful in restricted applications in such centres.

1.3.2 Inductive Electric Road Systems (I-ERS or DWPT)

The Dynamic Wireless Power Transfer (DWPT) used by I-ERS has the principle of resonant magnetic coupling as the basis [10]. It is done by inserting primary coils under the surface of the road; the current passed through these coils is of high frequency and of alternating current that induces a strong magnetic field. This field is reactive in the near-field zone and therefore generates a current in the secondary coil or pickup coil, which is located under the car and thus passes on the energy.

Operating Principles and Coil Topologies

To achieve maximum transfer of power and also minimise the reactive power requirement of the utility grid, the system needs to attain electrical resonance. This is normally achieved through the use of capacitors in series with the transmitting (primary) and receiving (secondary) coils, therefore, cancelling the network inductive reactance [12]. They are usually run at high frequencies, with reported frequencies of about 20 kHz [8, 13].

The top priority in the design of the coil pads is the geometrical design that maintains the efficiency of power transfer when operating in dynamic conditions and high speeds. Traditional coil shapes may be very intolerant of misalignment. As a result, new complex coil designs have been created and are extensively used in electric vehicles, including the Double-D (DD) coil design [12]. The reason behind the preference of DD coils over other simpler topologies is that these coils exhibit a high coupling strength and very low leakage flux [12]. This technology is vital, and it is important to reduce the leakage flux to minimise the energy loss and constant coupling coefficient of the vehicle during motion.

Challenges of Dynamic Misalignment and System Complexity

The main concern of I-ERS is to control the stability of power transfer during the motion of the vehicle. The misalignment between the primary coils embedded on the road and the secondary one mounted on the vehicle varies both laterally and longitudinally, as the vehicle passes by, dynamically and rapidly [8]. This unceasing change of relative position has a direct effect on mutual inductance and, therefore, on the coupling coefficient, which can cause significant decreases in the efficiency and reliability of charging [8]. New solutions, including the implementation of segmented transmitter coils that are only triggered when the vehicle is over the coils and their integration with active alignment monitoring systems, are necessary to reduce those momentary challenges and maintain constant power output [8].

The success of the I-ERS systems is solely dependent on the advanced topology of coils and active control systems that are highly responsive. This demand adds significant complexity to the power electronics and communications infrastructure of the system, adding significantly to the complexity of the systemic nature and cost density per kilometre as compared to the direct contact mechanisms of C-ERS. Moreover, the fact that resonant magnetic fields are going to be used in the transmission of power is also something that people have concerns about in terms of electromagnetic interference (EMI) as well as exposure of human beings to electromagnetic fields (EMF) [10]. The operation at high frequencies required to achieve efficiency in the transfer of power requires a stricter shielding solution and adherence to international standards of exposure to EMF. Vehicular and public safety ensures thus requires a holistic regulatory framework and focus on constant monitoring, which is another aspect of technical and policy challenges that were not experienced in C-ERS implementation.

1.3.3 Capacitive Electric Road Systems (CAP-ERS or CPT)

A technically feasible but less commercially developed variant of wireless power transfer is called capacitive power transfer (CPT), which uses resonant electric fields between parallel conductive plates [10]. The working principle is similar to a plate capacitor in which one of the sets of metal plates is integrated into the road and the other set of plates is integrated into the receiver, which is placed underneath the vehicle, with the road structure and airgap between being the dielectric medium.

Operating Principles and System Characteristics

The transfer of power in CPT is dependent on the maximisation of the resonance of the electric field. The road plates are excited by a high-frequency source, and the energy is picked up by the vehicle receiving plates. CPT has a number of theoretical benefits, such as the removal of eddy-current losses which plague inductive power transfer systems, and it can be easily worked around metallic devices [10]. More importantly, CPT systems have been shown to manage high power levels; there have been laboratory test systems with high-power systems that can transfer high power levels of up to 174 kW [10].

Constraints and Technological Readiness

The main engineering limitation that constrains the current practicability of conductive power transfer (CPT) to implement the deployment of dynamic electric road systems (ERS) is that it is inherently sensitive to the inter-component gap distance. CPT can only be used in a power transfer that is short, and is normally approximately a few millimetres to several centimetres [10, 14]. The high-power transfer needed to sustain this small gap requires extremely high operating voltages, and thus creates major issues regarding the safety, reliability of insulation and power-electronics design, especially when the infrastructure is open to the elements and debris. Keeping the height of a small, narrow air gap is very important to ensure consistent operation and necessitates built-in, dependable sensor systems to detect displacement as reported in parallel capacitive non-contact sensing studies [15]. The necessity of a very small, regular air gap is generally incompatible with the realities of the public road infrastructure, which is subject to much variation between surface wear, suspension dynamics of the vehicles, and environmental variation. This severe limit on gap distance essentially restricts the dynamic capability of CPT except in the case of advanced, expensive systems (e.g., active suspension devices on the vehicle or very smooth and flat road surfaces) that are added. This means that the current level of CPT is at an order of magnitude lower in Technological Readiness Level (TRL 3-5) than inductive power transfer (IPT) in a generalised dynamic road, and that current research is being dominated by power electronics optimisation and insulation durability in high-voltage environments.

1.3.4 *Comparative Analysis of ERS Technologies*

Strategic planning for national ERS deployment requires a robust comparative analysis of the three technologies, balancing the high efficiency and maturity of C-ERS against the seamless integration and safety benefits offered by W-ERS solutions.

Metric	Conductive ERS (catenary/rail)	Inductive ERS (IPT)	Capacitive ERS (CPT)
Power transfer mechanism	Physical contact (pantograph/rail collector)	Resonant magnetic coupling	Resonant electric field coupling
Typical efficiency (roadway-to-vehicle)	Very high (95%+) [9]	Medium–high (85–92%)	Medium–high (80–90%)
Tolerance to lateral misalignment	Low (strict alignment required)	Medium (improved by DD/bipolar coils) [12]	Low (highly sensitive to plate alignment)
Tolerance to air gap/ distance	N/A (contact required)	Medium (typically 10–20 cm gap)	Low (limited to few cm for high power) [10, 14]
Infrastructure visibility/ aesthetics	High (overhead structures/cables)	Low (infrastructure buried beneath road)	Low (plates integrated into road surface)
Maximum power density (kW/m)	Extremely high (proven in rail)	High (limited by coupling/heat)	Very high (potential for 174 kW+) [10]
Technological readiness level (TRL)	High (mature for rail/trolley) [3, 9]	Medium–high (pilot phase for dynamic road use)	Low-medium (focus of current research)

In conductive electric road systems (C-ERS), the greatest efficiency is provided through direct contact, minimising any transmission losses. Nevertheless, they are expensive in terms of establishing overhead structures (poles, gantries) and are opposed by society in terms of aesthetics. Inductive electric road systems (I-ERS) and CPT involve the use of expensive high-frequency, complicated power electronics within the roadbed and cost much more on a per-kilometre basis, but have the advantage that the core components are isolated behind the surface. Usually, ERS is more adaptable to be used in a wide range of vehicle categories and road geometry with tolerating high air gaps (10–20 cm). CPT has an extremely high potential power density and eddy current loss mitigation, but is severely limited by the need to have a small, narrow air gap [10, 14].

The analysis establishes that none of the ERS technologies are universal and best suited to all applications. C-ERS has proven itself to be more effective in high-volume dedicated heavy-duty vehicle (HDV) logistics corridors because of its capabilities and power-delivery potential [3, 9]. IPT is more adapted to the flexible and integrated deployment of urban and passenger vehicles in which aesthetic integration is important. The resulting diversity requires infrastructure investment strategies to have a mixed-technology design to suit individual traffic profiles, vehicle types and aesthetic needs. The creation of vehicle-side multi-technology compatibility standards is, therefore, the key to the achievement of commercial viability on the national level.

1.4 Benefits and Potential Impacts

1.4.1 Environmental and Energy Impact

The environmental justification of ERS goes beyond the core advantage of simply substituting internal combustion engines; it goes to the root cause of systemic inefficiencies of large-battery battery-electric vehicle (BEV) designs.

Emissions Reduction and Energy Efficiency

The ability of ERS to save on the size of the onboard battery is directly proportional to the savings in mass of the vehicle at large. As mentioned above, such mass reduction results in the saving of energy in the 13–17% in energy use per kilometre [5]. This high increase in the operational energy efficiency augments the overall effectiveness of electric mobility to the environment, by reducing the level of energy consumption needed to conduct transportation activities.

Renewable Energy Integration and Grid Stability

ERS is very important in the integration of high penetrations of intermittent renewable energy sources (RES). The ERS will be able to dynamically respond to the changes in the supply of the RES, turning the whole electrified road network into an enormous, geographically distributed energy consumer, which will subsequently change power consumption. As it has been established, the general decarbonisation effect of EVs is directly linked with the share of RES in the energy grid [1]. ERS helps in stabilisation of grid demand through load spreading, thus maximising the use of the renewable electricity produced when it is generated.

Reduction in Critical Raw Material Demand

The production of battery packs of large size necessitates large amounts of the raw materials that are essential, such as lithium, cobalt, and nickel. Ecological expenses incurred in the process of extracting, processing and recycling these materials are high. ERS can be used to dramatically reduce the aggregate demand of these materials, which is essential, due to the ability to right-size batteries, thereby satisfying supply-chain risk, geopolitical dependence, and overall environmental impact of EV manufacturing is significantly reduced.

1.4.2 Economic Benefits and Life-Cycle Cost (LCC) Reduction

The final economic feasibility of ERS will be determined by the fact that it will allow vehicle owners and fleet operators to achieve savings that will be greater than the high initial cost of the infrastructure development. This is mostly done by ensuring that the life-cycle cost (LCC) of the vehicle is significantly reduced.

Lowering Vehicle Purchase Price

The battery pack that is onboard is usually the most costly single part of an EV. ERS enables manufacturers to save on the installed battery capacity necessary to sustain a specific driving mission by a very large margin [2]. Such a direct downsizing of the battery size, in turn, keeps the initial purchase price of the vehicle downward [5], which is a strong incentive to increase the pace of consumers buying the vehicle in larger groups [4].

Fleet Management and Operational Cost Savings

In the case of commercial fleets, including logistics providers and mass transit organisations, ERS provides a 24 h operation, which removes the unproductive off time that was used in the past to charge the stationary charges. This radical growth in the utilisation of assets is directly transformed into better logistics efficiency. Moreover, through a decrease in the use of big batteries, ERS has the effect of greatly decreasing the cost of operation in the long term of the battery replacement and degradation procedures. As calendar ageing has been quoted as the leading degradation influence, representing up to 67% of the degradation expenses [7], unwanted capacity reduction is the most direct way to decrease this inherent, unavoidable capital cost throughout the service life of the vehicle of 10–20 years.

Table 1.1 Economic trade-offs: battery size reduction versus ERS infrastructure cost

Parameter	BEV (large battery)	ERS-enabled EV (small battery)	Impact of ERS deployment
Vehicle purchase price	High (battery cost dominant)	Low (due to battery right-sizing) [5]	Accelerates EV adoption; enhances market equity
Vehicle operational efficiency	Lower (due to mass penalty)	Higher (13–17% energy saving) [5]	Reduces LCC over lifetime
Battery life cycle cost (degradation)	High (calendar aging often dominant) [7]	Lower (reduced capacity/ less extreme cycling)	Extends service life of onboard batteries
Charging infrastructure requirement	High (static fast chargers and grid upgrades) [6]	Shifted to dynamic infrastructure (ERS) [6]	Reduces localized utility expenditure; manages peak load
Utilization of battery capacity	Low (large buffers often unused) [7]	High (optimized for buffer storage)	Improves capital efficiency

The following table summarises the key economic trade-offs facilitated by ERS deployment (Table 1.1).

1.4.3 Operational and Societal Impact

Enterprise-Level Energy Supply (ERS) generates significant externalities in mobility systems and urban planning. ERS enables continuous charging, ensuring that commercial fleets can operate without scheduled intervals. This results in increased productivity and reduced logistical bottlenecks that are essential to time-sensitive supply chains. In the crowded urban areas, the rollout of high concentrations of high-power fixed charging stations is competing for the limited real estate. The subsurface Intermodal Energy Supply (I-EST) system, specifically, introduces the concept of power transmission into the current roadway framework without much ado, decreasing the need to spread the specialisation of static charging stations. This assimilation increases the movement of mobility in the urban areas and frees the real estate to be used in other ways. Finally, by lowering the entry cost of the electric vehicles (EVs) and ensuring the minimum range tests on all members of society, ERS can support more widespread adoption of zero-emission mobility options, without limiting their adoption by high-income first adopters of new technologies.

1.5 Challenges and Barriers

1.5.1 Technical and Engineering Challenges

Scaling ERS infrastructure to cover vast distances requires addressing several formidable technical and engineering hurdles related to durability, efficiency, and safety.

Efficiency Maintenance Under Dynamic Conditions

In the case of wireless systems like inductive power transfer (IPT) and capacitive power transfer (CPT), the first and foremost engineering issue is to have high coupling efficiency when the vehicles are moving at high speeds and the positional relationships vary significantly due to lateral misalignment and varying vertical air gaps. High-speed frequency and power transfer adjustment of resonant frequency and stabilisation has to be used using advanced inverter and rectifier designs as the vehicle crosses charging segments on the millisecond scale [8].

Infrastructure Durability and Integration

The ERS elements incorporated into the roadway should be designed to resist serious, constant mechanical loads, as a result of heavy traffic loads, cyclic temperature fluctuations, and environmental exposures such as chemical assaults by de-icing agents. With buried Wireless-ERS (W-ERS), the stability and durability of the infrastructure depends solely on the attainment of strong ingress protection to coils and power electronics which are sensitive. In the case of Cable-Based ERS (C-ERS), despite the intrinsic robustness, the pantograph-catenary system needs continuous maintenance to guarantee the stable operation in the state of the real traffic load, but present data sets are not sufficient.

Electromagnetic Field (EMF) Safety and Mitigation

The functioning of wireless ERS invariably produces magnetic fields (through IPT) or electric fields (through CPT) [10]. Compliance with the specific requirements of international standards of safe use of EMF on the general populace and passengers is a condition that must not be compromised. This will require the creation and introduction of special shielding technologies that will be part of the roadway design and the vehicle chassis, providing additional tiers of cost and technical complexity to the system overall.

1.5.2 Economic, Regulatory, and Policy Challenges

The deployment of ERS on a national or continental scale requires capital investment measured in the hundreds of billions of dollars, demanding solutions to economic and regulatory complexities.

Massive Upfront Infrastructure Cost

The infrastructure of the necessary civil engineering and the high-density power electronics capital to power large road networks is an awe-inspiring initial capital cost. The provision of practicable funding mechanisms is the key to feasibility. Some of the possible models here are: public–private partnerships (PPPs), regulatory directives that make utilities invest in the ERS transmission level infrastructure, or a method of imposing advanced road-usage fees on the ERS users.

Lack of International Standardization and Interoperability

The main threat of large-scale implementation is that there is no international and unified agreement regarding the most essential parameters of ERS standardisation, including the operational frequency, the coil geometry, the communication protocols, and the payment schemes. Such regulatory gaps pose a threat to the development of anarchy, proprietary national networks, and thereby, the risk of increased redundancy, pushing the systemic costs above the appropriate level, and creating an almost complete impediment to cross-border logistics and freight movement [16]. Standardisation efforts by organisations like the Society of Automotive Engineers (SAE) and the International Organisation for Standardisation (ISO) are of critical importance, as they are the precondition of full-scale financial de-risking and market entry. The problems that have been experienced in the specification of interoperability requirements in other complicated areas of technological procurement like health information systems, highlight the difficulty in obtaining multi-stakeholder consensus in the context of ERS [17].

Regulatory Hurdles and Utility Integration

The implementation of ERS requires a way to navigate through difficult regulatory environments, such as infrastructure ownership, the right-of-way acquisition, and pricing framework decisions. Incorporation of the massive, mobile electrical load that ERS poses into the current utility regulation framework and planning paradigm has been a tremendous burden in bureaucratic and technical terms. The policy research should deal with the knowledge gaps and frequently conflicting interests between major stakeholders, such as transportation agencies, utility companies, and commercial providers of technology [17]. Besides, the experience in international pilot projects proves that effective policy integration, as well as strong local governance, is no less important to the project's success than technical feasibility. Institutional fragmentation, the lack of policy coherence and coordinated governance, may therefore raise the main implementation obstacle in most parts of the world, as opposed to the technical challenge that no one can overcome.

1.5.3 Social Acceptance and Adoption Barriers

In order to achieve the sustainability of ERS projects in the long term, societal acceptance and take into consideration equity in accessibility. Although overhead cable-based ERS (C-ERS) is highly efficient, it is often heavily opposed by the general population in some areas due to its perceived visual encroachment, as well as noise, and as such is often only applied within strictly delimited industrial or dedicated freight corridors. More so, whenever ERS infrastructure is funded by special user charges or dynamic charging methods applied on roads, policymakers should be prepared to respond in advance to issues that relate to equity and social justice. The design of policy frameworks should make the difference between vehicles accessing ERS charging and vehicles that cannot afford or are incapable of implementing ERS technology unintentionally create a two-tiered mobility system where vehicles that cannot afford or cannot implement ERS technology are economically or functionally disadvantaged.

1.6 Global and Regional ERS Initiatives

The shift of the electrified rail system (ERS) out of theoretical research to the industry reality is marked by the increase in the number of pilot projects worldwide, which introduce invaluable real-life information and policy lessons.

1.6.1 Major ERS Deployments in Europe and North America

Europe

The European countries have been the pioneers of massive, vibrant ERS pilots. Sweden. The broadly integrated e-highway project has studied the viability of conductive (ground rail) and inductive ERS systems in freight corridors in terms of complicated system integration and compatibility with other technologies. Overhead catenary C-ERS is also used in the German eHighway Project, but is mainly aimed at heavy truck platooning, and has demonstrated high operational efficiency and reliability of mature conductive technology in specialised logistics operations [3].

North America

University research centres and state transportation departments have been motivated in North America to conduct research and pilot projects, which have generally focused on I-ERS applications, especially in urban public transit buses and airport shuttles. The focus of these initiatives is on the smooth integration and exploitation of the existing utility infrastructure planning, normally done so through the public–private research alliances.

1.6.2 Focus on Asia: Korea's Commercial Trials and Policy Initiatives

South Korea has taken a progressive step in commercialising wireless ERS. The introduction of the Online Electric Vehicle (OLEV) technology, led by the organisation of KAUST, effectively implemented the quasi-dynamic I-ERS on the buses of the mass transit in cities. Such projects indicate the functionality and the great incorporation of wireless systems in the high-density urban spaces. Lessons on crucial policies that have been learned by South Koreans and especially in local government involvement, are that good governance and differentiation of local policy are essential in attaining sustainable and environmental success, and that local failure of poorly performing local environment arose out of a lack of integrated governance [18].

1.6.3 Policy and Research Acceleration in India and the Broader Asian Context

The wider Asian region is optimally developing the adoption of sustainable mobility solutions. One particular instance is the India-Japan Clean Energy Partnership, which is a strategic partnership scheme that has been initiated to jointly encourage the decarbonisation efforts of the two countries through pilot projects and research and development. Although such alliances frequently concentrate on green fuels (e.g. hydrogen), the policy impulse behind these alliances is the synchronous promotion of resilient, large-scale, and sustainable transportation infrastructure. Such international projects imply that ERS is seen as both a technological reaction to climate objectives as well as an important state-level economic development opportunity. Nations are positioning themselves well to position themselves as global technology leaders and dictate the future in the fast-paced industry of electrified logistics. The insights gained, including poor performance results in the cities of Korea, support the need to adapt policy on ERS to match the local energy systems, regulatory proficiencies, and distinctive urban planning issues [18] (Table 1.2).

Table 1.2 Summary of key global ERS pilot projects and governance lessons

Region/project	Primary technology type	Vehicle focus	Key technical outcome	Governance/policy lesson
Germany (eHighway)	Conductive (catenary) [3]	Freight/heavy duty trucking	High operational efficiency demonstrated (95%+) [9]	Validation of mature technology for dedicated, high-volume corridor use
Sweden (e-Highways)	Conductive and inductive mix	Freight/various	Tested multi-technology approaches and integration	Necessity of early standardization and public acceptance processes
South Korea (OLEV)	Inductive (DWPT)	Public Transit Buses	Successful commercial deployment in urban settings	Importance of strong local governance and integrated energy planning [18]
India/Japan (CE partnership)	Policy/R&D focus	Various (green mobility)	N/A (policy framework establishment)	Value of international strategic partnerships for technology acceleration

1.7 Research Gaps and Motivation for Study

Although pilot projects have promising outcomes, widespread implementation of ERS as a large-scale public infrastructure is hindered due to significant unanswered technical, economic, and policy challenges. These gaps in critical knowledge should be filled by thorough scholarly research.

1.7.1 Critical Technical Gaps

High-Speed Efficiency and Power Electronics

Whereas laboratory performance is satisfying, more research is urgently required in order to come up with robust, high-efficiency inverter and rectifier systems with high-power efficiency and long-life cycles in both thermal and mechanical conditions under continuous thermal and mechanical loading conditions in high vehicle speeds and under dynamic lateral misalignment.

Vehicle-to-Grid Integration with ERS

The potential of ERS implementation implies the possibility of EVs to offer supplementary services or to infuse power back into the grid. This functionality, coupled with a dynamically charging mechanism, needs completely new control policies and systemic modelling. The recent studies of the static show that profitability is immensely affected by the differentials of the electricity prices across regions and is commonly dominated by the battery ageing cost, especially calendar ageing [7]. ERS models need to determine how continuous, dynamic charging changes the battery state-of-charge profile and, in turn, how this diffuses or amplifies the various battery ageing factors during the lifetime of the vehicle.

1.7.2 Economic Gaps

Comprehensive Life-Cycle Cost (LCC) Modelling

The current economic models often do not reflect the holistic trade-offs that ERS implementation holds. There is a critical gap in creating strong LCC models that fully measure the total balance of costs: the huge investment in infrastructure should finance the cost savings in vehicles over a 20–30 year infrastructure horizon (credible and bankable cost–benefit analysis of the investment).

Infrastructure Cost Optimisation and Scalability

Studies are required to understand the most efficient density of ERS—i.e when a national road system is electrified, what proportion of the road system needs to be electrified to reach maximum utilitarian value (i.e. being able to support a small onboard battery to cover 95% of journeys) with minimum infrastructure investment? Such optimisation can only be done using sophisticated traffic modelling and spatial analysis, which are not yet standardised.

1.7.3 Policy and Governance Gaps

Developing International Interoperability Standards

Despite this recognised point of difficulty with standardisation [16, 17], formal, international standards of cross-technology and cross-border compatibility, especially of IPT geometry and communication, are evolving. The policy research should offer advice to the regulatory authorities, e.g. SAE and ISO, to strong, technology-neutral standards that de-risk investment and create large-scale competition in the market [16].

Sustainable Funding Mechanisms and Regulatory Frameworks

This large initial capital expenditure will require the creation of new and politically viable policy tools that have the ability to close the funding gap [4]. Further studies on effective governance patterns based on the pilot projects existing (e.g. South Korea) and innovative public–private financing patterns distributing the risks and rewards fairly are critical to the setting of precedents on adoption at large scales.

1.7.4 Justification for the Monograph

The reason behind writing this monograph is the pressing need to bring together the fragmented body of knowledge that exists between power electronics, civil engineering, economics and public policy into one unified, scholarly source. The rationale is to provide a narrow answer to the research gaps that are identified. The work will go beyond the qualitative comparisons and will provide the technical modelling of ERS operational dynamics in detail, with a specific focus on the issues of dynamic misalignment and efficiency maintenance. Combining these technical conclusions and technical–economic modelling (including LCC, battery degradation, utility planning, etc.), the monograph should produce practical, data-based advice on the strategies of implementing the national ERS, and assist researchers as well as guide the global policy-makers in creating the sustainable framework of the future, electric mobility.

1.8 Chapter Summary and Monograph Transition

In this introductory chapter Electric Road Systems (ERS) have become a conditioned next stage of electric mobility. It is driven by the global need to decarbonise and inherent systemic constraints, which are imposed by dependence on large-battery BEV systems, namely, economic, environmental, and operational constraints [5, 6]. ERS is a multi-purpose solution: it allows for right-sizing batteries, thus, at the same time, lowering the cost of the vehicle, making it more energy-efficient (13–17%) [5], and eliminating concentrated grid load. ERS is characterised, and the 3 major technologies outlined: Conductive ERS (high TRL, high efficiency, ideal HDV freight) [9]. Inductive ERS (flexible, medium TRL, congested by air-gap) 10 and Capacitive ERS (high power-density potential, low TRL, limited by air-gap constraint) [10, 14]. A mixed-technology deployment approach proves to be the best solution, as shown by comparison analysis. The transient requirement is, however, successful adoption, which is difficult due to critical technical issues, especially the sustenance of efficiency in the presence of dynamic misalignment, and institutional barriers associated with the massive funding of infrastructure and establishing non-proprietary international interoperability standards [16, 17]. As early experience with international programs—including the eHighways in Europe and the OLEV programs in South Korea—highlights, technology cannot be used without an efficient policy and governance framework [18]. This introduction forms the critical background of the following chapters, which will explore the necessity of the quantitative analysis needed to fill the identified knowledge gaps: Chap. 2 will shift the conceptual descriptions to mathematical modelling and optimization of the wireless power transfer, which in this case is inductive coupling systems and also discusses dynamic misalignment, frequency tuning, and coil geometry (DD, Bipolar) depending on the principles presented in Sect. 3.2. Chapter 3 will include detailed techno-economic models, performing the detailed Life-Cycle Cost (LCC) analysis of the ERS-enabled fleets, discussing the different models of infrastructure funding (PPPs, road charging) and overcoming the challenges associated with utility integration planning concerning the stability of loads and power delivery. Chap. 4 will explicitly review the worldwide regulatory and standardisation processes (SAE, ISO), examining the policy needs to gain the necessary interoperability and curb social and institutional obstacles to the prevalence of ERS.

References

1. Qian X, Zhong Y, Liu C, Zhang J (2024) The overall impact of EV adoption on CO_2 emissions is negative. Sci Rep 14:5785. https://doi.org/10.1038/s41598-024-55863-3
2. Setreus M et al (2017) The electric road system transition from a system to a system-of-systems. Procedia CIRP 60:185–190. https://doi.org/10.1016/j.procir.2017.01.019
3. Vantsevich V et al (2024) Electric road systems (ERS) for freight transport: enabling dynamic charging from the electric grid. Energies 17(3):1234. https://doi.org/10.3390/en17031234
4. Zavaleta J (2024) Achieving higher EV adoption in the U.S. and other lagging markets: addressing high costs, range anxiety, and charging infrastructure. World Electr Veh J 15(8):375. https://doi.org/10.3390/wevj15080375
5. Plötz P (2024) Impacts of increased battery size on BEV efficiency and energy use. Int Counc Clean Transp. https://theicct.org/publication/global-ev-battery-size-2024
6. Oladeji T et al (2024) Transformer load implications of large battery fast-charging in heavy-duty EVs. Energies 17(11):3011. https://doi.org/10.3390/en17113011
7. Ghorbanali H et al (2025) Calendar aging as the dominant degradation mechanism in BEV batteries. Sustainability 17(5):2456. https://doi.org/10.3390/su17052456
8. Sharma S (2021) Dynamic inductive charging efficiency under coil misalignment. J Phys Conf Ser 1950:012042. https://doi.org/10.1088/1742-6596/1950/1/012042
9. Fotopoulou D et al (2015) A survey of wireless power transfer and comparison of inductive vs capacitive coupling. IEEE Trans Power Electron 30(11):6017–6029. https://doi.org/10.1109/TPEL.2015.2427650
10. Mude V et al (2024) Enhanced coupling strength in DD coil topology for EV wireless charging. Energies 17(14):4220. https://doi.org/10.3390/en17144220
11. Kumar R et al (2021) Experimental prototype of inductive charging tested at 20 kHz. J Phys Conf Ser 1960:022045. https://doi.org/10.1088/1742-6596/1960/2/022045
12. Ma Z et al (2024) Air gap height measurement techniques for bearing-supported ERS. Proc Euspen
13. Nykvist B et al (2021) Benefits of integrating ERS as a charging option in BEV systems. Futur Transp 1(1):38–54. https://doi.org/10.3390/futuretransp1010004

14. Zavaleta J (2024) Interoperability for efficient EV charging ecosystems. World Electr Veh J 15(8):375. https://doi.org/10.3390/wevj15080375
15. Kim EY, Yoon JH (2024) Local government role in Korea's pilot emission trading scheme. Land 13(1):47. https://doi.org/10.3390/land13010047
16. Kalra K (2024) India–Japan clean energy partnership 2022: strategic initiative for decarbonisation. In: GH2 India report (2024) (Policy report, citeable as grey literature)
17. Muelaner J (2022) Electric road systems for dynamic charging (SAE research report EPR2022007). SAE International. https://doi.org/10.4271/EPR2022007
18. Escobar E, Carrillo A, María JLM (2021) Wireless power transfer technologies applied to electric vehicles: a review. Energies 14(6):1547. https://doi.org/10.3390/en14061547

Chapter 2
Global Technological Landscape of ERS

2.1 Introduction

Sustainable mobility is a key theme of the twenty-first century, with the pressing imperative to decarbonise the transport industry, the source of which is estimated at approximately one-sixth of the total emissions relating to energy [1]. Despite the battery electric vehicles (BEVs) adoption being a breakthrough, the use of a stationary charging infrastructure is a basic limitation to the complete electrification of road transport [2]. In the case of a large battery in a large truck, which is used in high-frequency modes of transportation, charging downtime and range anxiety are overwhelming obstacles to large battery sizes, making them heavy and reducing their capacity to carry a substantial load and be economical [3].

Electric Road Systems (ERS) or, to the layman, Charge As You Drive (CAYD) or dynamic charging are seen, in this case, as a disruptive infrastructure solution [2]. With the use of ERS technologies, vehicles are able to obtain electric power when in motion, and this has radically changed the way electric vehicles operate [4]. The ability to supply the necessary onboard battery capacity can be significantly lowered by ERS, which can supply constant energy onboard (up to 70% of the necessary capacity in some studies) [5]. The accumulated benefits of this reduction are immense: it makes EVs cheaper in the initial stage since battery can occupy 30–40% of the vehicle price [5]; it decreases the need of such raw materials as lithium, cobalt, and nickel [5]; it improves the efficiency of the vehicles by decreasing their mass and, consequently, the amount of wear on tires, brakes, and roads [6]. Furthermore, through the spread of load over the course of the day, ERS will be able to reduce the burden on the electrical grid commonly found when charging occurs in one place during the evening hours only [7].

The modern ERS movement is not just a technology, but a paradigm shift in the idea of electrification that has more of a vehicle-centric approach than a system-centric one. Traditionally, the development of EVs was centred on the vehicle aspect of the vehicle, and mostly this was improved by increasing the efficiency of batteries

R. Agnihotri, *Electrified Road Systems (ERS) for Emerging Asia*, Springer Tracts on Transportation and Traffic 24, https://doi.org/10.1007/978-3-032-16927-3_2

or the motor, but infrastructure was considered as an added part [8]. In opposition, ERS is also a symbiotic view of the vehicle and the roadway, in which the road ceases to be a passive surface and is transformed into an active energy delivery network [9]. This view comes about due to the understanding that an isolated vehicle optimisation, like the fitting of a truck with a huge battery to produce a specific range, is bound to give diminishing returns in terms of costs, efficiency, and sustainability of resources [2]. As a result, the present trend of international pilot projects represents a strategic turn on the issue of electrification at the infrastructure level, a turn with far-reaching consequences on grid management, urban planning, and policy.

2.1.1 Defining the ERS Technology Portfolio

The world of ERS has been referred to as three main technological archetypes, each having specific mechanisms, benefits, and issues that make each of them applicable in a variety of applications. It is necessary to briefly describe these technologies before going into a more technical analysis.

1. **Overhead Conductive Systems**: In this method, overhead catenary lines, similar to those found in trains and trolleybuses, are used as a source of power to vehicles. Tall vehicles (trucks and buses) have a pantograph mounted on the roof of the vehicle and make physical contact with the lines to draw electricity. It is an established and solid technology that is based on the existing and stable rail industry [2].
2. **Ground-Level Conductive Systems**: In this type of system, the power is carried through either physical contact with conductor rails or strips that are embedded into the surface of the road or on the surface of the road. Cars will be equipped with a sliding collector under the car chassis, which moves with the rail. This method is not as aesthetic as overhead lines and can be used on a wider variety of types of vehicles, such as passenger cars [10].
3. **Inductive (Wireless) Dynamic Charging**: This is a technology that allows the transfer of power in an electromagnetic field even without contact. The coils, which are embedded in the road surface, produce a magnetic field that is picked up by a receiver coil on the underside of the vehicle. The technique is the most flexible because it does not need any physical contact, has no exposed elements, and cannot be seen on the surface thus increasing the resistance to environmental factors [2].

2.1.2 Historical Context and Evolution

The idea of directly driving vehicles off the infrastructure is not new; instead, it is an old concept that is being re-energised by the developments in power electronics, material science and the urgent necessity to reduce climate change felt in the world.

The history of the emergence of ERS can be dated back to the development of electric mobility. The concept of overhead conductive transfer of power was already proved in 1882 when Werner von Siemens tested his trolleybus in Berlin and used the technology to date, with more than 300 trolleybus systems still operating around the world as of 2018 [11]. On the same note, patents on inductive energy transfer were registered as early as the 1890s; at the same time, the electric car was initially gaining prominence [11].

First electric vehicles were popular in the decade turn of the twentieth century and were praised as being quiet, clean, and easy handling in comparison to those powered by gasoline and steam [8]. But the introduction of the mass-produced Ford Model in 1908, the discovery of massive amounts of cheap crude oil and the invention of extensive road systems saw the internal combustion engine's supremacy and the removal of the EV in a matter of decades [12].

The energy crises of the 1970s led to a renewed interest in electric propulsion by showing the weakness of fossil-fuel dependency. This was the time frame when the first ERS research began, such as the Roadway-Powered Electric Vehicle (RPEV) project of Lawrence Livermore National Laboratory in the early 1980s. In this project, the transfer of inductive energy to a modified Volkswagen was demonstrated in a modified test track (220 m) with the help of contemporary technology, thus showing that the concept is viable [12].

There is an interplay of factors that are driving the current global proliferation of ERS pilot projects. The need to attain a net-zero carbon emission level has developed strong policy impetus, which is coupled with the technological maturity of all three varieties of ERS, which have recently moved beyond a laboratory concept to practically demonstrable systems on public roads [13]. What is more, the chapter will examine the changing environment where stationary charging infrastructure is necessary, but not sufficient to decarbonise the whole road transport industry, especially the long-haul, freight sector.

2.2 ERS Technologies—Technical Overview

The engineering principles, operational specifications, and implicit trade-offs of the three major technology families will have to be analysed in detail to obtain a comprehensive picture of the situation in the global ERS environment. There is no single system that is more efficient, expensive, scalable, or resilient than others, and this determines the appropriateness of a system to various applications and environments.

2.2.1 Overhead Conductive Systems

The most technologically developed type of ERS is overhead conductive systems, often known as catenary systems or eHighways, which borrow decades of development experience in the electric rail and trolleybus sectors [11].

Mechanism: The infrastructure of a two-pole system is a series of roadside masts or poles that carry a two-pole overhead contact line system, and typically suspended some 5 m above the roadway to supply both the power in-feed and the necessary DC voltage to the road vehicular load [14]. In contrast to a single-pole system in use in railways, which makes use of the rails as a return, a two-pole system is required to supply the needed DC voltage and the power in-feed to serve the road vehicles.

The focus of vehicle adaptation is an active or intelligent pantograph, which is fitted on the roof of the truck or bus. This is much more advanced than a basic collector; it uses sensors to connect to and disconnect from the overhead lines when the vehicle is moving, reaching speeds of up to 90 km/h [14]. The active pantograph will dynamically regulate its connection to provide a stable connection even when the truck is laterally moved in its lane or the road height changes, and it automatically closes when the vehicle switches lanes, passes the electrified corridor or the battery is switched off.

Power Specifications and Efficiency: Catenary systems are unique in terms of their capability to carry a very high amount of power, thus making them very well adapted to heavy-duty transport. Pilot projects have proven power delivery of 500-650 kW with system designs capable of 1 MW [15]. It tends to utilise low DC voltages (670–750 V), which makes it the most efficient of all the ERS technologies in grid-to-wheel. Research and experiments exhibit an excellent efficiency of 80–85%, which is roughly twice that of the standard internal combustion engine, and the system is naturally a two-way system, so the regenerative braking energy can be returned to the grid, further improving the overall system efficiency [16] also, the system is two-way by nature, so the regenerative braking energy can be fed back into the grid, which is further increasing the overall system efficiency (Fig. 2.1).

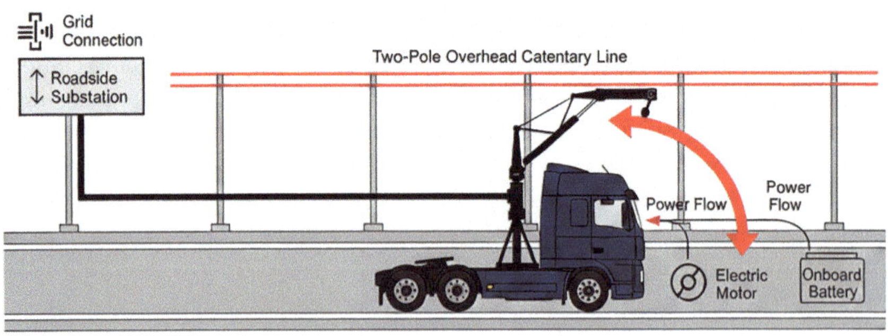

Fig. 2.1 Schematic of an overhead conductive ERS

Advantages

- **Power and Efficiency**: It is the first benefit of the system since it can provide high power with minimum loss, which is the key feature of the system that allows it to be the most successful in electrifying the heaviest vehicles on the busiest routes.
- **Ever-tested Technology**: The major elements rely on proven, dependable technology in the rail industry, minimising technological risks and tapping into the current supply networks and maintenance innovations [11].
- **Reduced Infrastructure Cost**: For inductive systems, the cost of catenary infrastructure is usually higher, with some reports showing a range of 1.7–3.1 million per direction per kilometre per direction.

Challenges

- **Visual Intrusion**: The masts and overhead wiring network have some considerable aesthetic effects on the landscape, which may provoke opposition in the populace, especially in urban or picturesque locations [9].
- **Poor Portability**: The system only works with the tall cars, including trucks and buses. Passenger cars are not practical and safe at all, as they would need impractically long pantographs to access the lines in addition to restricting the number of persons that can use them and the overall utilisation of the infrastructure [17].
- **Infrastructure Vulnerability**: The overhead lines are open to weather damage caused by extreme weather conditions, crashes involving high vehicles, and should be maintained regularly to maintain reliability.

2.2.2 Ground-Level Conductive Systems

Ground-level conductive systems are an intermediate between the high efficiency of conductive power transfer and the need to have smaller intrusive infrastructure. They are designed to be more universal than their overhead equivalents.

Mechanism: This technology entails a conductive rail, which is laid on the road infrastructure. The rail can be built out of the pavement, or it can be mounted on the surface; an important safety consideration is the division of the rail; it is subdivided into short, electrically isolated segments, each of not many metres in length. When a suitably equipped and approved vehicle is detected to be above a section, the section is energised, and when the vehicle has passed by, the section is de-energised within a few milliseconds. Such an on-demand solution, known as the live-on-demand, will help to avoid the threat of electric shock to people, animals, or other vehicles. Cars are modified by adding a mechanical collector, or pick-up, which is a retractable device installed on the undercarriage. This collector can be lowered to make physical, sliding contact with the energised rail, and thus, they can pass power to the electrical system of the vehicle, regardless of its size, such as passenger cars and heavy-duty trucks,

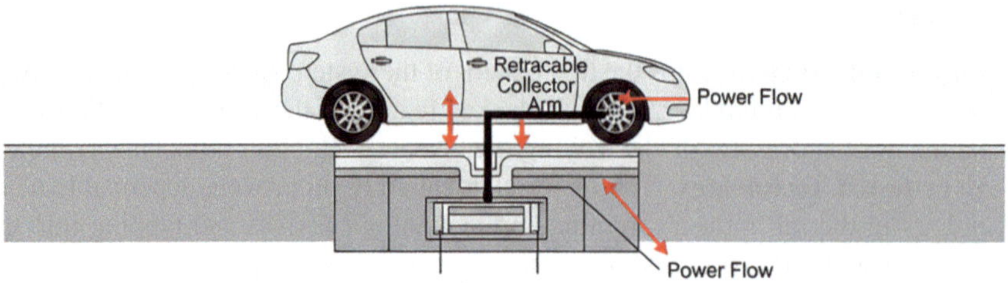

Fig. 2.2 Cross-section of a ground-level conductive ERS

as long as they have the necessary collector installed [17]. This technology can be used in all types of vehicles, including passenger cars and heavy-duty trucks.

Power Specifications and Efficiency: Ground-level conductive systems have proven to be able to transfer a large amount of power, which is applicable in a large host of vehicles. The transfer of power over 90 kW has been measured continuously with a potential of short peaks near 180 kW, which is adequate to propel and charge batteries of heavy vehicles [18]. The direct electrical contact provides remarkably high efficiencies in these systems. More recent simulation research, which has been tested on a physical demonstrator, has demonstrated end-to-end efficiencies of 96.5% in urban traffic and 95.7% in rural highway conditions [18] (Fig. 2.2).

Advantages

- **Universal Applicability**: Ground-level rails can support all forms of electric vehicle (unlike catenary systems), which means that the infrastructure can be utilised by a greater number of vehicles and that the market is expanded, which enhances the utilisation rate [19].
- **High Efficiency**: The system maintains the high efficiency of conductive power transfer that reduces energy losses between the grid and the vehicle to a minimum [18].
- **Minimal Visual Impact**: The visual impact is not expressed by the infrastructure that is laid in the road surface and does not affect the surrounding landscape of the environment, making it more appropriate in urban and scenic landscapes [17].

Challenges

- **Mechanical Wear and Maintenance**: Mechanical Wear and Maintenance: The mechanical wear between the collector and the rail is due to the direct physical contact of the two components, which requires periodic inspection and maintenance [20].
- **Environmental Susceptibility**: The rail will be subject to the road environment, and this can include debris, sand, water, ice and snow and may disrupt electrical connection or necessitate specialised road-maintenance, e.g. non-metallic snowplough blades [10].

- **Retrofitting Costs and Complexity**: Incorporating the rail system with the current road surfaces is a complicated and high-cost civil-engineering endeavour that may incur milling the road surface and may cause much traffic disruption. The rail can also pose a problem in future resurfacing of the roads [4].
- **Safety Concerns**: Although the segmented activation design is incorporated, the high-power electrical conductor in the public roadway poses safety concerns that have to be strictly met by well-established engineering and fail-safe control measures [10].

2.2.3 *Wireless Dynamic Charging (Inductive)*

Dynamic transfer of power is the most technologically advanced and flexible way of ERS, which is referred to as Inductive dynamic charging or Dynamic Wireless Power Transfer (DWPT), which does not require any physical contact between the car and the road.

Mechanism: It is a technology based on resonant magnetic coupling, similar to that used in wireless chargers of consumer electronics but scaled to high-power applications [21]. The infrastructure comprises a series of copper coils, which are primary transmitters, buried about 8–10 cm beneath the asphalt surface, connected to roadside management units, which tap power into the grid and convert it to a time-varying magnetic field [22].

A coil of secondary receiver is placed under the body of the vehicle as a pad. This coil is set to give the same frequency as the transmitter coils. There is a current induced in the receiver coil as the vehicle is travelling through the magnetic field. This AC electricity is then picked up by onboard power electronics in the vehicle, rectified into DC power and used to power the motor and charge the battery simultaneously, only when a compatible vehicle is in sight; thus, being safe and only consuming a small amount of standby power [23].

Power Specifications and Efficiency: Power transfer capabilities of inductive systems have continued to enhance with the continuous research and development. Protests have reached an average power transfer of 100,000 W to a 40-ton vehicle moving at 80 km/h [24]. When it comes to buses, rates of about 30,000 W have been dynamically demonstrated [25]. The grid-to-wheel efficiency is naturally less than that of conductive systems due to the losses incurred in generating the magnetic field and in the space between the coils. In optimal conditions, reported efficiencies are normally between 85 and 90% [26] (Fig. 2.3).

Advantages

- **Sturdiness/Low Maintenance**: Due to the completely subsurface nature of the system, it is immune to the effects of meteorological factors like snow, ice and rain, therefore, overcoming the vulnerability of the exposed components to mechanical erosion and wear and tear.

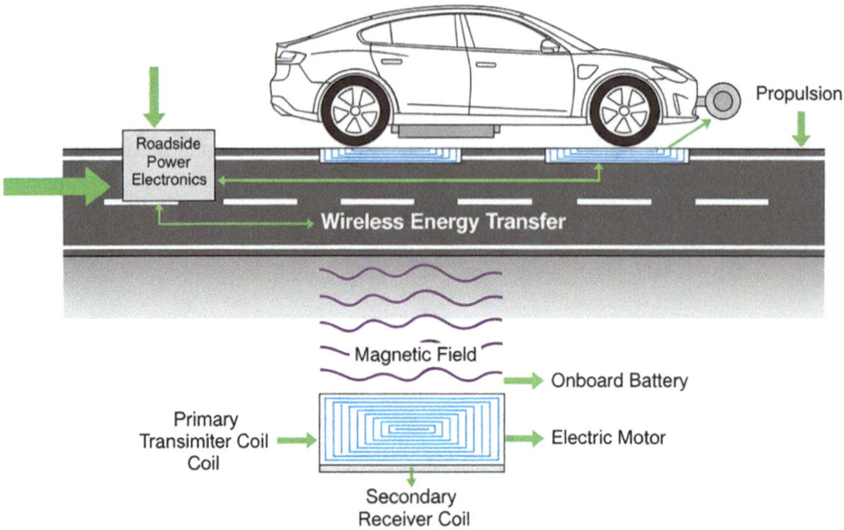

Fig. 2.3 Principle of inductive (wireless) dynamic charging

- **Natural Safety and Beauty**: The full burial of the infrastructure provides a higher degree of safety for all road users by removing exposed live parts of electricity, and it gives zero aesthetic visual effect to the surrounding scenery.
- **Universal and Flexible**: The technology applies to all electrified vehicles such a passenger cars, buses, and trucks and can facilitate autonomous vehicle operations due to its nature of being automated.

Challenges

- **Reduced Efficiency**: Converting energy (electricity to a magnetic field and vice versa) is less efficient compared to direct conductive power technology, making the process less energy-efficient.
- **Alignment Sensitivity**: The effectiveness of power transfer depends upon the tightness of both the lateral and vertical position of the transmit and receive coils; a large amount of misalignment reduces energy transfer to the vehicle.
- **High Infrastructure Cost**: A typical characteristic of inductive electric road systems is that they involve the highest capital expenditure because they require sophisticated power electronics in each road segment and a complicated installation procedure.
- **Electromagnetic Field (EMF) Exposure**: Concerns persist about EMF exposure among the public and the government. The systems are designed to meet high global safety standards (e.g., ICNIRP guidelines), and exposure is alleviated via a segmented activation strategy, but this is one of the focal points where the public needs to be informed and justified.

These technologies have a complicated engineering trade-off between maximising efficiency and ensuring long-term operational resilience. The most efficient grid-to-wheel efficiency is offered by conductive systems, overhead or ground-level systems that create a direct electrical circuit. These advantages come, however, with a high level of primary vulnerability, as mechanical elements are prone to wear and environmental hindrances like debris and ice, which increase the level of maintenance. On the other hand, the inductive systems tolerate a relative loss of efficiency in favour of high resilience. The system provides the solution by moving energy across an air gap through a magnetic field to remove moving parts in the system and seal the active infrastructure with the environment and leading to remarkable weather and wear resistance. This basic trade-off implies that the best technology choice is not universal but exceedingly dependent on the particular application, climatic conditions, traffic density and tolerance to maintenance of the operator compared to capital investment (Table 2.1).

Table 2.1 Comparative overview of ERS technologies

Feature	Overhead conductive	Ground-level conductive	Inductive (wireless)
Power transfer mechanism	Direct contact via overhead catenary lines and vehicle pantograph	Direct contact via in-road/on-road rail and vehicle collector	Resonant magnetic coupling between coils in road and vehicle
Typical power delivery	500–650 kW; suitable for heavy-duty vehicles	>90 kW continuous, 180 kW peak; suitable for all vehicle types	30–100 kW; suitable for all vehicle types
Grid-to-wheel efficiency	80–85% (highest)	95–97% (highest)	85–90% (lower)
Vehicle adaptation	Roof-mounted active pantograph	Undercarriage retractable collector	Undercarriage receiver coil and power electronics
Key advantages	Mature technology, highest power transfer, relatively lower cost	Universal vehicle applicability, high efficiency, low visual impact	No mechanical wear, weatherproof, inherently safe, zero visual impact
Key challenges	High visual impact, limited to tall vehicles, infrastructure vulnerability	Mechanical wear, susceptible to debris/ice, complex retrofitting, safety of exposed rail	Lower efficiency, alignment sensitivity, high capital cost, EMF concerns
Est. cost/km (both directions)	€1.7–3.1 million	€1.2–2.7 million	€1.2–2.7 million

2.3 Global Pilot Projects

Deployments in the real world explain the theoretical capabilities and the challenges associated with the electric road system technologies in a better way. In the past ten years, various pilot projects have been launched across the world, led by governments and also by industries to test, validate and demonstrate these systems under working conditions. The empirical information on such projects is priceless in terms of performance, cost, acceptance among the population and scalability. In this section, the analysis of four of the most important pilot initiatives will be given, where each of them is a specific technological and strategic approach toward the implementation of the electric road system.

2.3.1 Sweden—Smart Road Gotland (Electreon)

Sweden has become a leading country in electric roads research, with a national goal of net-zero by 2045 and 2,000 km of electric roads by 2030. The Smart road Gotland project is one of the demonstrations of the country. Technology and Scope: The project was pioneered by the Israeli technology firm Electreon and installed a 1.65 km inductive wireless electric road system on a portion of public road between Visby and the Gotland airport. The pre-commercial demonstration was aimed at developing the knowledge and evaluating whether the large-scale wireless transmission of energy in the form of traffic could be transferred into reality.

Vehicles and Power: The two vehicle types that were used to test the system were a fully electric 40-ton heavy-duty truck with a trailer and an electric bus that serves the airport shuttle. One of the successes was the fact that the shared charging platform was demonstrated as capable of charging two vehicles at the same time, despite the different battery systems and power demands. During the testing, the system was able to transfer an average power of 100 kW into the truck as the truck reached a speed of up to 80 km [1].

Funding Model: The project was designed as a public–private partnership with the Swedish Transport Administration (Trafikverket) being the leading financier that granted the tender to the Smart Road Gotland consortium, headed by Electreon AB. The Swedish Energy Agency and the European Regional Development Fund were also additional stakeholders in providing funds. A one-year extension with upgrading of a 400 m stretch of road was supported by an extra budget of EUR 2 million by Trafikverket, which highlighted the interest of the government in the research phase.

Conclusions and Final Results: The Smart road Gotland project confirmed the technical feasibility as well as operational strength of inductive electric road systems. The critical conditions proved that the technology can last long under harsh Nordic winter conditions; it was confirmed that the system could be used efficiently and operated even at −23 °C without deteriorating the charging performance. Independent structural analysis by the Swedish National Road and Transport Research Institute (VTI) found that the installation of charging coils did not have any negative impact on the structural stability and life span of the road. Nevertheless, with these technical achievements, the project demonstrated that there are major economic imbalances to scalability. Though Sweden had an envision of a larger and permanent electric road on the E20 highway, the procurement procedure was cancelled due to bids received by possible suppliers that were higher than the budget of the project. In a later final report by the wider Swedish electric road systems programme in 2024, it was suggested that the idea of implementing a national electric road network was not the way forward at the current time, with it being deemed that cost-efficiency would only be met once the technology has been widely adopted by other major trading partners like Germany and France. This conclusion reveals the most important lesson that technical feasibility does not necessarily mean economic viability on a national level.

2.3.2 Germany—Electreon Trials

After Sweden, Germany has initiated some of the key pilot projects to test Electreon's sliding-rail energy-retrieval system (ERS), where inductive technology in both public transport and highway use is mainly emphasised. These initiatives form part of a comprehensive national plan that is to be used to decarbonise the transport sector.

Projects and Scope: Two salient German trials are using the technology of Electreon that are being conducted:

- **ELINA (Balingen)**: This experiment uses a 1 km inductive ERS in a non-privacy road within the municipality of Balingen and also supplements the system with two fixed wireless charge stations located at bus stops. The purpose of this is to recharge an electric bus that is going to travel a local route. The programme is held in partnership with energy utilities company EnBW as well as sponsored by the German Aerospace Centre (DLR) [27].
- **EMPower (Autobahn)**: The given initiative is going to build a 1 km wireless ERS on a stretch of the Bavarian Autobahn. It is subsidised by the German Federal Ministry of Economic Affairs and Climate Action (BMWK). Its main aim is to test its whole ERS value chain, including planning and construction as well as operation; it also tries to examine its viability in a heterogeneous fleet of buses, trucks, and passenger cars in high-speed highway conditions [28].

Funding Model: German ERS projects are usually intended to be government-funded research projects that feature public–private partnership (PPP) consortia consisting of technology providers, industrial companies, and academic institutions. Within the ELINA project in Balingen, the Electreon project to install the charging infrastructure is funded to the tune of EUR 3.2 million as part of the broader programme of the federal government to fund innovative e-mobility initiatives, the Elektro-Mobil programme [28, 29].

Outcomes and Challenges: The first outcomes of the Balingen trial are encouraging. With the joint application of dynamic en-route charging and static opportunity charging at stops, it is possible to make the electric bus work across days and still have no net drop in its battery state of charge (SOC). This result shows the possibility to avoid the need to charge battery overnight, therefore, allowing use of smaller, lighter, and cheaper batteries and, thus, decreasing the overall cost of ownership to the public transport operator.

However, the implementation of ERS in Germany is facing a challenge that goes beyond technical reasons. The energy industry in Germany is extremely decentralised, consisting of more than 800 separate distribution grid companies, which complicates the integration of the grid and its approval by the regulators, as well as the financial guarantee of the active support of mass infrastructure projects such as ERS [30]. Recent changes in policy, including the sudden withdrawal of electric-vehicle purchase subsidies at the end of 2023, have created a significant level of market uncertainty and question the political will to support an aggressive e-mobility transition, which may pose a threat to funding and support of large-scale infrastructure projects like ERS.

2.3.3 United States—California eHighway (Siemens)

Unlike the European willingness towards inductive technology, the major ERS demonstration in the United States was to use a more advanced overhead catenary system to meet one definite and powerful environmental challenge.

Technology and Scope: In a partnership between Siemens and the South Coast Air Quality Management District (SCAQMD), the eHighway California project led to the installation of a one-mile (1.6 km) overhead conductive system in a major traffic route of the south-Alameda Street in Carson, California. The site has been strategically located because it is close to the Ports of Los Angeles and Long Beach, where heavy-duty drayage trucks are the major contributors to the smog-causing emissions [31].

Vehicles and Power: The display included three separate trucks with different hybrid powertrain setups: a battery-electric truck, a diesel-hybrid truck, and a compressed natural-gas (CNG) hybrid truck [31]. This strategy underscored the flexibility of the technology to work with a wide variety of onboard energy sources and allow trucks to operate without emissions under the electrified tracks and then turn to their corresponding hybrid systems when exiting the electrified tracks. High efficiency of the system was highlighted, and Siemens recorded that the system is twice as efficient as the standard diesel engine [16].

Funding Model: The project was a multi-stakeholder PPP and had a total budget of USD 13.5 million. The money was raised through various sources, including USD 2.5 million provided by SCAQMD, USD 4 million by a legal settlement with a shipping company, USD 3 million by the California Energy Commission and USD 2 million each by both the Port of Long Beach and the Los Angeles County Metropolitan Transportation Authority. The technology partner, Siemens, also supplied an extra USD 1.3 million in its in-kind [31].

Outcomes and Expansion Plans: The principal result of the eHighway demonstration was the successful, technically speaking, validation of the zero-emission cargo transport system in a real-world urban freight corridor [31]. The project resulted in the confirmation of the fact that the catenary system could be successfully used to supply heavy trucks, which would allow reducing the number of local air pollutants, decreasing the consumption of fossil-based products, and lowering the cost of running the fleet [16]. Nonetheless, the successful completion of the pilot project did not lead to any tangible plans of expanding the catenary ERS on a large Following state-level transportation and energy reports have been inclined to focus on the development of stationary fast-charging facilities for both passenger and heavy-duty vehicles [32].

2.3.4 South Korea—Gyeonggi Province Test Bed (KAIST OLEV)

South Korea has a tradition of being a leading country in terms of commercialising electric rolling stock (ERS) and showing the first public bus service in the world, which was based on the dynamic use of wireless charging. Although this showed a technical breakthrough, the ultimate fate of the program highlights the essence of looking into non-technical barriers to ERS implementation.

Technology and Scope: The Korea Advanced Institute of Science and Technology (KAIST) created the Online Electric Vehicle (OLEV) system, which was opened in 2013 in the city of Gumi, with infrastructure at 5–15% of the total route, mostly at stops and selected points, required to enable the buses to have adequate battery charge [26].

Power and Vehicles: OLEV buses were designed to take power at 100 kW with a carrier frequency of 20 kHz, with a reported transmission efficiency of 85% with a 17 cm air gap between roadway and vehicle [33]. This has allowed the use of batteries one-third the size of those required by a standard depot-charged battery-electric bus.

Funding and Cost: The programme was mainly funded by the South Korean government [34]. The technology was found to be costly, and this was a significant obstacle to wider use. OLEV buses were reportedly going to cost about US $630,000 each, which is a huge premium when compared to the conventional diesel buses [35].

Outcomes and Lessons Learned: The Gumi OLEV project was a significant technical success, and this proved that the idea of using dynamic wireless charging could be effectively used to serve a commercial public transit service. The advantages of minimised battery sizes and the removal of downtimes were well displayed. However, the project turns out to be a warning story; after a few years of usage, the bus service in Gumi and other locations was already stopped, and the authorities explained the decision by the old infrastructure and the scandal surrounding the further financing of the technology by the government [11]. Sustainable business model, maintenance and capital replacement plan, and long-term political and social support of the way it is operated are the elements that cannot be left out of the long run success.

The mutual experience based on these global pilots indicates varying strategic frameworks of ERS implementation. The European projects in Sweden and Germany are the expression of a "Universal Public Good paradigm, in which the inductive technology is the focus, and the overall view is a universal charging network that is available to all types of vehicles, which makes the focus on the universal applicability and aesthetic inclusion in the world of the public." The pilot of the United States in California, on the other hand, stands as an example of a Targeted Industrial Solution model, making a pragmatic choice of the most mature and powerful technology (catenary) to deal with a specific, high-impact, industrial and environmental problem, port drayage emissions, and accepting operational constraints because of the use case defining the application. Lastly, the first attempt of South Korea can be described as a Technology First model, which succeeded in reaching a world-first technical demonstration but failed because it lacked an economic and operational structure that was not yet developed. The long-term trend of ERS on a global scale is probably determined by which of these strategic paradigms, or a combination of any of them, will turn out to be economically sustainable and politically feasible in the long-term perspective (Table 2.2).

2.4 Comparative Analysis of ERS Technologies

Proper choice of an ERS technology requires a multidimensional evaluation, which takes into consideration rival points like energy saving, economic price, scalability, and safety. Practical evidence of the pilot projects conducted in other parts of the world offers the necessary information that pushes the debate towards one of theoretical specifications rather than actual performance. No one technology comes out as universal and better than the others, but each has a unique set of strengths and weaknesses that fit particular strategic priorities.

Table 2.2 Summary of global ERS pilot projects

Country	Project name	Technology	Project length	Vehicle types	Power delivered	Key outcomes	Funding model and cost
Sweden	Smartroad Gotland	Inductive (wireless)	1.65 km	40-ton truck, bus	100 kW (to truck)	Validated tech durability in winter; no negative road impact; deemed not cost-effective for national scale-up at present	PPP; funded by Swedish transport admin (€2 M for extension) and others
Germany	ELINA/E	MPOWER	Inductive (wireless)	1 km (each)	Bus, truck, car	~30 kW (bus)	Demonstrated ability to eliminate need for depot charging for buses; testing highway feasibility
United States	CA eHighway	Overhead conductive	1.6 km (1 mile)	Hybrid trucks (BEV, diesel, CNG)	Not specified, but high power for HDVs	Proved feasibility for zero-emission drayage; reduced local emissions and operating costs	Multi-stakeholder PPP; $13.5 Million total budget
South Korea	KAIST OLEV	Inductive (wireless)	24 km route (5–15% electrified)	Public buses	100 kW	World's first commercial wireless charging bus service; enabled smaller batteries. Service discontinued due to funding/maintenance issues	Government-funded; high vehicle cost (~$630 k per bus)

2.4.1 Energy Efficiency

Energy efficiency, which is measured as a ratio of grid-to-wheel power transfer, is a very important measure considering that it directly impacts the cost of operation and general benefit to the environment. In this field, such hierarchy is evident.

- **Ground-Level Conductive Systems** have the greatest possible efficiency. At direct sliding electrical contact, and with such a small number of conversion stages, such systems can be made with efficiencies above 95%, along with resistive dissipation in the rail and collector, and losses in semiconductor loss in the power electronics to segmented activation that is exaggerated by the stop-and-go urban traffic environment [18].
- **Overhead Conductive Systems'** efficiency is also high. The systems, utilising mature technology in the use of the railway, always have grid-to-wheel efficiencies in the 80–85% range and can regenerate braking energy and recycle it to the grid.
- **Inductive (Wireless) Systems** is the least efficient of the three types. It normally attains efficiencies between 85 and 90%, with the most significant loss resulting from the energy required to generate the magnetic field and the imperfections inherent in the air gap between the roadway and the vehicle. Efficiency is sensitive to factors such as vehicle velocity, the physical air-gap distance, and, most importantly, the lateral positioning of the vehicle receiver over the transmitter coils of the road [36]. Although improvements have been made in coil design and power electronics, there is still no fundamental efficiency advantage over conductive methods.

2.4.2 Installation and Maintenance Costs

Economic feasibility of ERS is largely dependent on the initial cost of installation (CapEx) and the operational expense (OpEx) of maintenance in the long run. The cost estimates are highly heterogeneous because the technology is new, and the conditions in the area are different, but we could outline certain tendencies.

- **Installation Cost (CapEx)**: Cost estimates of any of the technologies are highly uncertain but typically lie within the range of EUR 1.2–EUR 3.1 million per kilometre to electrify a road in both directions. Traditionally, overhead catenary systems were believed to be less expensive in terms of investment than in-road systems [14]. However, these costs are coming together, as in-road technologies are growing. Inductive systems are normally the most expensive in the startup cost since a large amount of power electronics is needed to handle each of the coil segments and the civil works needed to place the coils into the pavement [14]. Ground-based conductive systems also have a lot of road work, and this can be disruptive and costly, especially in urban retrofit systems. Overhead systems do not cause significant disruption to the road surface, but demand a considerable amount of masts and sub-stations.

- **Maintenance Cost (OpEx)**: It is at this point that the technologies are greatly differentiated. Inductive systems are the least costly in the long run in terms of their maintenance. Having no moving components and all working components covered with asphalt, they are not vulnerable to mechanical wear and environmental degradation [21]. The roadside power electronics are mostly limited to maintenance. Conductive systems (overhead and ground level), on the other hand, are more expensive to maintain because they are a mechanical wear and tear system. The catenary wire or road rail and the pantograph or collector of the vehicle are prone to friction and are inspected and cleaned regularly, as well as replaced periodically [20]. The ground-level rail is especially susceptible to road wear, road debris and snow-removal activities and may result in increased lifetime expenses [21].

2.4.3 Scalability and Application Suitability

ERS technology can be scaled based on the flexibility of the technology to various roads (urban vs. highway) and the type of vehicles (light vs. heavy-duty), as well as the possibilities of widespread implementation.

- **Overhead Conductive**: It is a very scaled-up technology that is targeted at dedicated freight corridors. It has a high power capacity, which is why it is good to electrify large highways with high truck traffic, as done in Germany and experimented with in California [37]. Nonetheless, its size is inherently small in that it cannot serve passenger vehicles and is also massive in appearance, rendering it inappropriate to be used in large quantities in cities or scenic areas [14]. It is not a general-purpose solution, but rather a niche solution to heavy-duty arteries.
- **Ground-Level Conductive**: This type of technology is highly scalable to vehicles of all sizes since it can be used on a small passenger vehicle up to 40-ton trucks [19]. This is what renders it an attractive alternative in mixed-traffic settings, urban and rural. Nevertheless, the scalability is subject to the logistical complexity and the cost of retrofitting the current road networks and also the long-term durability issues of mechanical wear and exposure to the elements [20].
- **Inductive (Wireless)**: Inductive ERS potentially has the highest scalability opportunities on a broad and long-term basis. Its ability to fit all forms of vehicles and the fact that it has a low visual impact and is very strong can be used in any virtual environment and includes old city centres as well as major highways [9]. The biggest obstacle to its scalability is now economic, namely, the high initial capital cost [14]. Nevertheless, once the technology matures and production is increased in scale, the cost is likely to reduce, which could result in it becoming the most diverse and universally adoptable solution in the long term.

2.4.4 Safety and Environmental Impact

The key to any public infrastructure project is its safety and environmental concerns:

- **Safety**: The three systems are all designed with high regard with respect to safety. Inductive systems can be assumed to be the safest since no visible high-voltage components are present [22]. The issue of EMF exposure is mitigated by designing systems to meet international health requirements and through segmented activation, where coils are only energised when a vehicle is in motion [26]. Conductive systems pose more of an engineering challenge in the safety aspect. Ground-level systems are based on ultra-fast, fail-safe segmented operation to avoid electric shock, one of the most vital systems that has to be perfectly reliable [13]. Although overhead systems are based on technology successfully deployed in the rail industry, physical infrastructure (masts) is introduced, which might become a collision risk; downed lines under extreme weather conditions or accidents should be dealt with.
- **Environmental Impact**: The main environmental advantage of all ERS: the first one is the lessening of greenhouse gas emissions and local air pollution caused by road transport. Though the technologies themselves have varying environmental footprints. The masts and wires used in overhead systems have the greatest visual and land-use effect [17]. Conductive and inductive in-road systems have a smaller physical footprint, but necessitate large quantities of energy and materials to construct and retrofit the road. There is also an environmental cost of manufacturing the components, which are: copper coils, power electronics, and steel rails. The complete environmental impact of every system, including manufacturing and installation, operation and decommissioning, requires a complete life-cycle analysis (LCA) to compare the two systems.

To conclude, the selection of an ERS technology is a strategic decision, which implies the prioritisation of these competing factors. In a high-volume, high-dedication freight corridor where power and efficiency are of top priority and visual consideration is of secondary importance, overhead catenary systems are a strong option. In-road systems can be chosen better in situations of mixed-use urban or highway networks where universal access and aesthetic integration are important. At that classification, the choice should be between conductive and inductive technology; here lies a trade-off between the efficiency and low CapEx of conductive rails, and resilience, reduced maintenance and intrinsic safety of inductive charging.

2.5 Funding Models

Electric Road Systems (ERS) implementation is a large-scale infrastructure project, and the costs involved in capital implementation are several million euros per kilometre. This way, the development of feasible and sustainable funding schemes is as vital as the technological development itself. Pilot projects in the international arena provide a new experience of the existing financial systems, which are largely based on public-sector leadership and collaborative projects.

2.5.1 Public–Private Partnerships (PPPs)

The Public–Private Partnership (PPP) model has emerged as the most common model of funding and implementing ERS pilot projects. This model has the advantage that the capital and regulatory controls are provided predominantly by the public administrations (including national transport ministries or energy agencies) whilst technological know-how, project management, and even a share of co-investment is provided by the private enterprises. Sweden (Smart road Gotland) provides a good example: this project, launched by the Swedish Transport Administration (Trafikverket) and other public funds, has provided a tender to a consortium headed by a technology company, Electreon, which is privately owned. This system allows the government to promote national agendas such as decarbonisation and leverage the innovation and flexibility of the business domain.

- **Sweden's (Smart road Gotland)**: provides a good example: this project, launched by the Swedish Transport Administration (Trafikverket) and other public funds, has provided a tender to a consortium headed by a technology company, Electreon, which is privately owned. This system allows the government to promote national agendas such as decarbonisation and leverage the innovation and flexibility of the business domain.
- **Germany's Electreon trials**: The trials of Electreon in Germany follow a similar tendency. ELINA and EMPower, among others, are supported by the government research and innovation programmes (via the Deutsches Zentrum für Luft- und Raumfahrt (DLR) and the Federal Ministry of Economy and the Environment (BMWK), respectively), and the consortia include private companies (Electreon, EnBW, Volkswagen), infrastructure firms (Eurovia), and academic institutions that carry out the work.
- The California eHighway project can be seen as another multi-stakeholder PPP; however, it has a less homogeneous funding coalition. Public funds include the integration of federal, state, and regional organisations (SCAQMD, California Energy Commission, LA Metro, Port of Long Beach) with legal settlement and also an in-kind donation by the technology partner, Siemens. It is an illustration of a model according to which several public organisations with similar interests in the form of freight mobility and air quality can invest in the project as co-founders.

The PPP model is favourable during the demonstration stage since it allows the governments to take the excessive financial risk of undemonstrated technologies, at the same time encouraging the innovation process. However, the question of the long-term scalability of this model is one of the main ones.

2.5.2 *Government Grants and Public Funding*

The primary stimulus of the present-day ERS ecosystem is direct government funding, mostly in the form of grants and special infrastructure schemes. The emergence of ERS has been mainly motivated by societal requirements and policy goals as opposed to direct market demand, thus making public subsidies inevitable.

The Bipartisan Infrastructure Law (BIL) includes programmes funded by the federal government such as the National Electric Vehicle Infrastructure (NEVI) Formula Programme (US$5 billion) and the Discretionary Grant Programme on Charging and Fuelling Infrastructure, (US $2.5 billion) in the United States, which are large federal funding pools that can potentially be exploited to fund ERS projects, even though the latter currently focuses on non-portable charging. The U.S. Department of Transportation provides programmes as well—e.g. the Innovative Finance and Asset Concession Grant Programme—to aid in the exploration of the PPPs in transportation projects.

Renowned national governments and the European Union fund research and innovation in sustainable transport in Europe. An example of such grants is the so-called Elektro-Mobil programme by the German government, which forms the basis of the EM Power project, and which is necessary to de-risk the technology and build the knowledge base needed to commercialise it [28].

2.5.3 *Private Investment and Commercial Models*

Though the situation is dominated by the current prevalence of public funding, the long-term vision of the ERS depends on the ability to attract major investment from the stakeholders. This will necessitate that there is creation of viable business models that can yield a satisfactory return on investment (ROI). Several possible commercial models are currently being investigated:

- **Charging as a Service (CaaS)**: Technology providers, like Electreon, are offering a CaaS model by which the company funds the initial capital investment of infrastructure and retrofits of vehicles. The fleet operators then charge a subscription fee or even charge per kilowatt-hour of usage, similar to utility bills. This method will swap the huge initial CapEx to a foreseeable OpEx to the user, hence improving the chances of adoption.
- **Toll-Based Models**: In the case of ERS installed on large highways, a tolling system might be implemented where the users will pay a fee based on the amount of electricity they use during a passage through the electrified lanes. This would work just like the current road tolls, where the revenue will be used to meet the operating and capital expenses of the infrastructure. Based on the telecommunications sector, revenue management systems would be needed to handle metering, billing, and dynamic pricing on a load basis, depending on the grid load [34, 38].

- **Asset Concession Models**: In this form of PPP, a long-term concession is awarded to a private party to design, finance, operate and maintain an ERS corridor. The concessionaire will then make a profit by charging users. Models in support of this model include the U.S. DOT programme, the Innovative Finance and Asset Concession Grant Programme [39].

2.5.4 *Cost-per-Kilometre Analysis*

A cost per kilometre analysis is required to determine the economic viability of ERS. However, in the current circumstances, with the pilot products, there is a great level of uncertainty in these figures.

- **Infrastructure Costs**: Estimates of the costs of installing ERS in both directions of a road incurred on either side amount to approximately EUR 1.2 to EUR 3.1 million per kilometre. The wide variability of cost–benefit analysis was demonstrated in a research in the Netherlands, which used a range of EUR 10 million to EUR 5.0 million per kilometre in its analysis [7, 40].
- **Total Cost of Ownership (TCO)**: The TCO is a central issue as viewed by the user. Research has shown that battery electric vehicles (BEVs) equipped with ERS can have a TCO edge over conventional diesel trucks and compete with battery electric vehicles that are not powered by stationary chargers, at least in the case of heavy-duty trucks. The main factors that affect TCO are the cost of electricity (in the case of ERS, depot charging, and public fast charging), the cost of the battery, and scheme of finance of the infrastructure [41].
- **Cost–Benefit Analysis (CBA)**: Socially, a complete CBA should be carried out, whereby all direct investment and operation costs will be put in addition to external advantages like a decrease in greenhouse gas emissions, clean air, and reliance on imported fossil fuels. Research has also indicated that user fees might not be sufficient to cover the cost of the whole investment, but overall benefits would justify the implementation of ERS in the concerned sectors [42].

The business case of Electrified road Systems (ERS) is expected to entail the shift of the current paradigm of 100% government funding of demonstration projects towards the hybrid frameworks combining the pools of public funds, capital investment by the companies, and the sources of revenue generated by the users. Such a transition will be possible not only due to the empirical demonstration of the technical reliability of ERS but also because it will demonstrate the ability to offer a strong economic case to fleet operators, as well as it will be able to offer a quantifiable societal payoff.

2.6 Challenges and Considerations

Despite promising technological advancements and successful pilot demonstrations, the widespread deployment of Electric Road Systems faces a complex array of technical, economic, and social challenges. Overcoming these hurdles is critical for ERS to transition from niche pilots to a scalable component of the future transportation infrastructure.

2.6.1 Technical Challenges

Despite the proven principles of ERS technologies, their reliability, efficiency and safety functioning at the scale of large-scale and real-life functioning remain to be a challenge in terms of the technical aspects.

- **Durability and Maintenance**: In the case of in-road systems, whether conductive or inductive, the issue of long-term durability of embedded components is a major issue. Infrastructure should have the capacity to endure sustained traffic loading, temperature variations and exposure to moisture, and mechanical forces of road maintenance operations such as resurfacing [9]. The Smart Road Gotland project has shown that inductive coils can resist road forces and extreme weather conditions over a long period of time [25]. Although this performance has yet to be proved over several decades. The conductive systems are another challenge that presents a mechanical wear to the rail and collector, which requires a durable and economical maintenance program [20].
- **Efficiency and Alignment**: Structurally resilient, inductive systems, however, have to deal with reduced energy efficiency compared to conductive methods. This performance is very delicate to the spatial correlation between a vehicle receiver and roadway transmitter coils [13]. This performance is a significant engineering difficulty in a heterogeneous fleet of vehicles dynamically travelling in the road. This misalignment results in reduced transfer of power, and this may compromise the system's effectiveness and economic justification.
- **Interoperability and Standardisation**: The relevant issue of scalability is the existing absence of standardisation between differing ERS technologies and vendors. To operate in a similar way as a utility, the vehicles manufactured by any manufacturer should be able to feed off any ERS infrastructure. The demands of this need include internationally accepted standards of communication protocols, frequency of power transfer (in inductive systems) and the physical connector (in conductive systems). Without such standards, the market would be at risk of being divided into proprietary, incompatible systems, which would make vehicle manufacturers and infrastructure developers disinterested in investing in the market [11].
- **Electromagnetic Field (EMF) Exposure and Interference**: In the case of inductive systems, the safety of vehicle occupants, pedestrians, and people having

medical implants must be ensured regarding the safety of EMF. Although the systems in place have been designed to meet the present international safety standards 36, the perception and regulatory appraisal of people involves constant confirmation and open lines of communication. Additionally, the possible threat of electromagnetic interference (EMI) with other car systems or roadside communication systems should be effectively reduced [11].

2.6.2 Economic Challenges

Economic factors are the major challenge to the implementation of ERS. The initial capital expenditure on the infrastructure is high, and this poses a dreadful economic challenge.

- **High Capital Cost and ROI**: Established cost of installing a national highway system: installation expenses of a national highway network would cost tens or hundreds of billions of euros, which is a substantial amount of money that governments must raise funds for when faced with competing priorities. The ROI is long-term and spread among society, through environmental gains and users, through lower operating costs, making it not easy to justify it based on the simple business case of the commercial businesses [42]. As seen in Sweden, even with the technical success of a project, the decreased costs of large-scale deployment are too high to justify it just on a simple business case basis.
- **The "Chicken-and-Egg" Problem**: There is a standard problem of coordination that seals off ERS implementation. The investors in infrastructure are afraid to build electric roads without taking into consideration the number of vehicles with ERS to be able to generate revenue. On the other hand, the automobile makers and fleet operators are hesitant to invest in ERS-compatible vehicles unless they are sure that there will be a large-scale, cross-interoperable charging system. The only way out of this dilemma is a bold decision by the government and primary funding by the citizens to install a base network and provide an initial market push [35].
- **Competition with Alternative Technologies**: ERS has not developed in a vacuum. It has to compete with the fast time-evolving environment of fixed-charge technologies and alternative fuels. The improvement of battery technology to make them cheaper, higher energy densities, and shorter charging times may make ERS less advantageous in some applications [4]. And another possible decarbonization route is the creation of hydrogen fuel cell technology in long-haul trucking. These competing solutions will require the ERS to be able to provide a more attractive total cost of ownership and operational model that will determine its long-term viability.

2.6.3 Social and Regulatory Challenges

In addition to technical and economic conditions, success will be impossible without social approval and a favourable regulatory environment.

- **Public Acceptance**: The social attitude can be a determining factor of ERS. The overhead catenary systems, though efficient, are often met with resistance by people because of the perceived aesthetic disruption caused by their appearance in the setting [17]. In-road systems also require extensive construction, which can cause long-term traffic jams, creating resentment amongst commuters and the people living in the area. Public concerns, either scientifically based or not, about the safety of in-road conductors or the EMF emitted by wireless systems may also be a reason why this technology should be accepted by the whole community, and the effects of construction on the surrounding community should be kept to a minimum.
- **Regulatory and Institutional Frameworks**: ERS deployment crosses the conventional realms of transportation, energy and urban planning, thus creating complex regulatory issues. The following are the key questions to be answered: Who runs the ERS infrastructure, the road authority, a utility company or a private concessionaire? How will the electricity be billed and priced to the users? What will the connection be with the electricity grid, and who will be responsible for making necessary grid upgrades? The creation of legal and regulatory frameworks that would clearly define these matters is a precondition to attracting private investment and allowing large-scale deployment [38].
- **Equity and Accessibility**: With the implementation of electric road systems (ERS), the issue of distributive justice emerges. A two-level transportation system with electric roads being implemented as the high-tiered tolled lanes might result in one of the two outcomes: the environmental and logistical benefits of electric mobility can only be applied to those who can afford the toll fee or the necessary vehicle retrofits. Policymakers require that mechanisms be evaluated that would allow the benefits of ERS to be widely spread across society, and not become concentrated in the hands of the already advantaged in society, including large economic regions, as well as in underserved areas and rural locations.

The case of the South Korean orthogonal linear electric vehicle (OLEV) project is a very good example of these challenges. Although a technical success, the eventual unsustainability of the project was due to a combination of high cost, a long-term necessity to continue to be financed by the government and to be affected by the economic, social and political factors that led to its long-term success [11]. This demonstrates the need to take a holistic approach that not only covers the engineering factors but also the economic, social and political factors that would ensure the success of the electric road systems in the long term.

2.7 Future Outlook and Research Gaps

The evolution of the global ERS scene is in the pace of lightning speed, and continuous research and development of the field tries to overcome the dominating impediments and open new abilities. Technological innovation, greater integration with an emerging energy ecosystem, and the creation of friendly policy frameworks are going to be the drivers of ERS in the future. The next step of the ERS development will follow several essential tendencies and areas of research gaps.

2.7.1 Emerging Technologies and Efficiency Improvements

Ongoing improvement of the ERS technology is aimed at improving efficiency and cost reduction, as well as power density.

- **Hybrid ERS Systems**: Hybrid ERS systems are now discussed as a future implementation, where, in larger highway segments, high-power overhead catenary networks will be implemented, and in cities or tunnels, where visual impact is an issue, inductive systems or conductive systems on the ground are implemented. It is a hybrid solution that is supposed to bring the best performance and maximise the acceptance of the people in different settings.
- **Wireless Efficiency and Power**: Research in the field of wireless efficiency and power transfer aims at reducing the difference in efficiency compared with conductive systems and increasing the level of power transfer. The investigations include enhanced coil material and geometry, high-frequency power electronics utilising wide-bandgap semiconductors, including silicon carbide (SiC) and gallium nitride (GaN), and advanced control algorithms that can adapt to dynamic alignment scenarios [36].
- **Advanced Materials and Construction**: Material science is currently being advanced in order to come up with more durable and affordable parts of the ERS infrastructure. These involve wear-resistant alloys in conductive railings and collectors, and composite materials that are created to provide the rapid installation of the inductive coils with a minimum level of disruption on the roads. It is also looking into modular construction methods, which can greatly help in cutting down installation time as well as the cost associated with it [43].

2.7.2 Integration with Smart Grids and Renewable Energy

Probably the most radical opportunity that ERS has is its combination with smart grid architecture and renewable energy sources. Energy ERS is not just a transportation technology, but it is a distributed energy infrastructure.

- **ERS as a Grid Asset**: Smart ERS infrastructure as a grid asset can serve the role of a flexible load, balancing the grid by dynamically changing charging requirements in real time, in reaction to grid conditions. As an example, the charging power can be slightly decreased when peak system-wide demand is low and slightly increased when there is an excess of renewable energy [44].
- **Synergy with Renewable Energy**: ERS specifically fits very well as a supplement to intermittent renewable energy sources, which include solar and wind energy. Large battery arrays and roadside energy storage systems can be co-located with ERS infrastructure to provide storage of excess renewable generation that can subsequently be discharged to power the ERS in case of high demand or low renewable output [45]. This co-location approach increases system resilience, increases its efficiency, eliminates the need to upgrade the grid, and optimises the use of clean energy in transportation [46].
- **Vehicle-to-Grid (V2G) Integration**: ERS might facilitate a two-way energy flow by vehicle-to-grid (V2G) in the long term. A battery located in a vehicle may serve as an ancillary grid services when running on the road, like frequency regulation. This would turn electric vehicle fleets into active participants in the energy market as well as passive consumers despite the enormous technical and regulatory challenges that are still present [44].

2.7.3 Scalability Pathways and Policy Frameworks

To transform ERS to a large-scale deployment, clear strategy routes and support in terms of policy are required. The strategies of targeted deployment must be focused on the high-impact corridors, instead of trying to achieve a national coverage at the initial level.

- **Targeted Deployment Strategies**: Initial large-scale deployments are unlikely to cover entire national road networks. Instead, a more strategic approach is needed, focusing on high-impact corridors. These include:
  - **Urban Public Transport Routes**: High-frequency bus corridors can be electrified to realise fast decarbonization of urban transport and enhance urban air quality, and allow a smaller and cheaper battery [1].
  - **Major Freight Corridors**: By focusing on key logistics corridors between ports, distribution centres, and industrial hubs, it is possible to reach a disproportionate number of emissions associated with the heavy-duty trucking industry [37].
  - **Last-Mile Delivery Hubs**: Implementing inductive charging infrastructure at depots and delivery paths will improve operational efficiency in the growing e-commerce industry [17].
- **Need for Policy and Standardisation**: Governments are significant in ensuring that conditions are favourable towards ERS scale-up. This includes:

– **Establishing a Long-Term Vision**: A national roadmap. The clear and long-term roadmap is a guarantee that the industry will invest in technology and manufacturing [47].
– **Driving Standardisation**: The state government's efforts to make international interoperability standards are essential to avoiding market fragmentation [11].
– **Providing Financial Incentives**: Ongoing government subsidies, tax credits, and creative financing schemes, including green bonds, are needed to close the economic gap and lure private funds at the early stages of large-scale implementation [48].

2.7.4 Key Research Gaps

In spite of the considerable progress, there are still a number of research gaps to be filled in so as to guide the further development of the Electric Road Systems (ERS) and guide policy makers.

- **Long-Term Performance and Life-Cycle Analysis**: Pilot projects may offer useful short-term empirical data, but there is a strong desire to have longitudinal research over more than a decade to determine the durability and maintenance needs and degradation processes of ERS infrastructure under realistic working conditions. Detailed life-cycle assessment (LCA) is also required to compare the overall environmental impact of the different ERS technologies, including the extraction of materials all the way up to decommissioning.
- **Socio-Economic Impact and Business Models**: More studies need to be done to better optimise business models and to measure the more global socio-economic consequences of ERS implementation. This will include in-depth total cost of ownership (TCO) studies of each of the fleet types, cost–benefit analysis (CBAs) that incorporate the benefits of reduced pollution to the population, and studies of possible impacts on employment and economic growth.
- **Grid Impact at Scale**: Since the effect of single pilot projects in the electric grid can be confined, the effect of a larger-scale, multi-thousand-kilometre ERS network on national and regional power systems is not well understood. Simulations and modelling projects on a large scale are required to design sufficient grid reinforcements and develop complex energy management plans.

The convergence, the integration of advanced materials, the intelligent grid technologies, and the active public policy are likely to be the future of ERS. Although major obstacles are still in place, the possibility of ERS becoming a foundation of a truly sustainable, efficient, and stable transportation system can hardly be disputed. This potential will only be realized through further research, strategic investment and international partnership.

2.8 Conclusion

In this chapter, the author has outlined the overall technological environment in the world of Electric Road Systems, tracing their development from a historical idea to a real-life example. The discussion shows a dynamic and lively sphere of innovations, which is motivated by the common global goal of decarbonization of road transport. ERS technologies overhead, conductive, ground-level conductive and inductive wireless, each offer a unique array of technical functionality, economic trade-offs and strategic implementation. There is no single best solution, but the best solution depends on the use case scenario, the environmental situation and the long-term policy goals.

A case study on the major pilot projects in Sweden, Germany, the United States and South Korea has provided important lessons. All three types of technologies have undergone strong technical viability, which confirms their ability to minimise the use of large batteries, remove range anxiety, and reduce emissions. But these projects have also placed in a very dramatic way the non-technical barriers to adoption, which are so formidable. The prohibitive nature of infrastructure capital, the requirement of sustainability of long-term funding models that extend past initial government grants, the difficulty in attaining interoperability by standardisation, and the need to have sustainable public and political backing have taken the form of the most pressing challenges.

The experience of Sweden and South Korea, especially, can be taken as warning examples, as it shows that even well-developed technological projects can fail in case the economic and political structures are quite strong. The problem-oriented, task-focused nature of the California eHighway, on the one hand, and the systematic, research-oriented strategy in Germany, on the other hand, depict different possibilities to go in.

In the future, the future of ERS is closely connected to the overall trends in energy and technology. The combination of ERS and smart grids, and renewable energy sources has the potential to make this technology not just a charging solution but a dynamic tool that will improve grid stability and use clean power to the fullest. The efficiency and cost reduction will continue with further innovation of materials and power electronics, and system design, which will make the economic case of deployment more robust.

Finally, the pilot projects for a scalable electric road network are not only a technical problem but a socio-technical change that is a complex process to make. It needs a very inclusive strategy that will correlate the technological innovation with good economic models, favourable policy structures and popular will. The variety of solutions that have been contemplated in this chapter regarding technologies and strategic approaches implies that there will probably be a collection of custom solutions, instead of a one-size-fits-all solution, when it comes to the future of electrified transport. The experience of these early endeavours will form an invaluable base for the policy-making or infrastructural investments that will define the next generation of sustainable mobility, to be discussed in more detail in later chapters of this book.

References

1. International Energy Agency (IEA). Electric Vehicles. https://www.iea.org/energy-system/transport/electric-vehicles. Accessed 1 Oct 2025
2. Akhilesh A. Exploring electric road systems (ERS): a pathway to sustainable mobility. https://www.akhilesh.info/exploring-electric-road-systems-ers-a-pathway-to-sustainable-mobility/. Accessed 1 Oct 2025
3. Conductix-Wampfler. Roadway power supply—eroad systems. https://www.conductix.com/en/solutions/roadway-power-supply. Accessed 1 Oct 2025
4. Electreon. Technology overview. https://electreon.com/technology/. Accessed 1 Oct 2025
5. Transport & Environment. Electric road systems for heavy-duty vehicles. https://www.transportenvironment.org/discover/electric-road-systems-for-hdvs/. Accessed 1 Oct 2025
6. Chalmers University of Technology. Dynamic charging of electric vehicles on roads. https://www.chalmers.se/en/areas-of-advance/energy/strategic-areas/e-mobility/electric-roads/. Accessed 1 Oct 2025
7. Trafikverket (Swedish Transport Administration). Electric roads. https://www.trafikverket.se/en/startpage/operations/Operations-road/Electric-roads/. Accessed 1 Oct 2025
8. Swedish Transport Administration. Roadmap for electric road systems 2020–2030. https://www.trafikverket.se/contentassets/roadmap-ers.pdf. Accessed 1 Oct 2025
9. European Road Transport Research Advisory Council (ERTRAC). Electrification roadmap. https://www.ertrac.org/. Accessed 1 Oct 2025
10. Energimyndigheten (Swedish Energy Agency). Electric roads. https://www.energimyndigheten.se/en/sustainability/electric-roads/. Accessed 1 Oct 2025
11. Volvo Group. Electric roads and heavy transport solutions. https://www.volvogroup.com/en/news-and-media.html. Accessed 1 Oct 2025
12. ABB. Electrification solutions for transport infrastructure. https://new.abb.com/ev-charging. Accessed 1 Oct 2025
13. Bureau of Energy Efficiency (BEE). Standards and guidelines for EV charging infrastructure. https://beeindia.gov.in/. Accessed 1 Oct 2025
14. Smartroad Gotland. Pilot electric road project in Sweden. https://www.smartroadgotland.com/. Accessed 1 Oct 2025
15. Japan Electric Vehicle Association (JEVA). On-road charging and power transfer studies. https://www.jeva.or.jp/. Accessed 1 Oct 2025
16. Korea Institute of Energy Research (KIER). Dynamic wireless power transfer for road vehicles. https://www.kier.re.kr/eng/sub03/sub03_04_01.html. Accessed 1 Oct 2025
17. MDPI. Benefits of an electric road system for battery electric vehicles. https://www.mdpi.com/2032-6653/13/11/197. Accessed 1 Oct 2025
18. Ministry of Road Transport and Highways (MoRTH), Government of India. National electric mobility mission plan (NEMMP). https://morth.nic.in/. Accessed 1 Oct 2025
19. ElectReon. Smart road infrastructure for dynamic charging. https://electreon.com/projects/. Accessed 1 Oct 2025
20. Central Electricity Authority (CEA). Charging infrastructure guidelines for electric vehicles. https://cea.nic.in/. Accessed 1 Oct 2025
21. Scania Group. Electric road technology development. https://www.scania.com/group/en/home/products-and-services/electric-road.html. Accessed 1 Oct 2025
22. International Transport Forum (ITF). Decarbonising transport: electric roads in future mobility. https://www.itf-oecd.org/. Accessed 1 Oct 2025
23. Tata Power. Public EV charging and smart grid integration in India. https://www.tatapower.com/. Accessed 1 Oct 2025
24. United Nations Economic Commission for Europe (UNECE). Transport electrification policies. https://unece.org/. Accessed 1 Oct 2025
25. World Economic Forum (WEF). The future of road electrification. https://www.weforum.org/. Accessed 1 Oct 2025

26. Shell Recharge. Electric road system pilots and global charging initiatives. https://www.shell.com/. Accessed 1 Oct 2025
27. Springer Nature. Advances in electric road system integration. https://link.springer.com/. Accessed 1 Oct 2025
28. Elsevier. Hybrid charging models for electrified transportation. https://www.elsevier.com/. Accessed 1 Oct 2025
29. Taylor & Francis Online. Infrastructure readiness for electric roads. https://www.tandfonline.com/. Accessed 1 Oct 2025
30. SAE International. Standards for wireless power transfer for road vehicles. https://www.sae.org/. Accessed 1 Oct 2025
31. Department for Transport, UK. Feasibility of electric road systems for the United Kingdom. https://assets.publishing.service.gov.uk/government/uploads/electric-road-system-feasibility.pdf. Accessed 1 Oct 2025
32. Green Car Congress. Electric road projects and pilot deployments. https://www.greencarcongress.com/. Accessed 1 Oct 2025
33. McKinsey & Company. The future of electric roads: a global outlook. https://www.mckinsey.com/. Accessed 1 Oct 2025
34. ResearchGate. A comparative analysis of static and dynamic wireless charging systems for electric vehicles. https://www.researchgate.net/publication/384139876. Accessed 1 Oct 2025
35. Nordic Energy Research. Electric road systems. https://www.nordicenergy.org/wp-content/uploads/2019/10/Electric-roads_web.pdf. Accessed 1 Oct 2025
36. PwC. Electric road systems: opportunities and challenges. https://www.pwc.com/. Accessed 1 Oct 2025
37. Road Research International. Wireless charging of electric vehicles: efficiency studies. https://www.researchgate.net/publication/344567892. Accessed 1 Oct 2025
38. Lund University. Electric road systems research. https://www.lu.se/electric-road-systems. Accessed 1 Oct 2025
39. MDPI. Reducing the environmental impact of large battery systems with conductive electric road systems—a technical overview. https://www.mdpi.com/2032-6653/15/2/59. Accessed 1 Oct 2025
40. Electreon. Gotland, Sweden electric road case study. https://electreon.com/projects/gotland. Accessed 1 Oct 2025
41. Electreon. Findings from the world's first public wireless electric road for heavy-duty commercial vehicles. https://electreon.com/articles/worlds-first-public-wireless-electric-road. Accessed 1 Oct 2025
42. Electreon. Electreon announces extension of world's first wireless electric road for trucks and buses. https://electreon.com/articles/electreon-announces-extension-of-worlds-first-wireless-electric-road-for-trucks-and-buses. Accessed 1 Oct 2025
43. World Bank. Sustainable electric mobility infrastructure development. https://www.worldbank.org/. Accessed 1 Oct 2025
44. California Energy Commission. Electrifying California's highways. https://www.energy.ca.gov/sites/default/files/2024-04/CEC-600-2024-017.pdf. Accessed 1 Oct 2025
45. MDPI. Conductive electric road localization and related vehicle power control. https://www.mdpi.com/2032-6653/13/1/22. Accessed 1 Oct 2025
46. Visão Zero 2030. Electric road systems: a solution for the future? https://visaozero2030.pt/wp-content/uploads/Electric_Road_Systems_solution_future.pdf. Accessed 1 Oct 2025
47. TNO. Electric road system: social cost benefit analysis. https://www.ttm.nl/wp-content/uploads/2025/08/tno-electric-road-system-social-cost-benefit-analysis.pdf. Accessed 1 Oct 2025
48. Fraunhofer Institute. The possible future of electric road systems in Europe—time to decide and act. https://publica.fraunhofer.de/bitstreams/7ac7fe00-408f-4467-a17b-92c41018fc49/download. Accessed 1 Oct 2025

Chapter 3
Dynamic Wireless Charging: Principles and Applications

3.1 Introduction

The process of worldwide transportation electrification is an essential component of policy development aimed at preventing climate change and improving the quality of air in cities. Nevertheless, the mass implementation of electric vehicles (EVs) is essentially limited by the drawbacks of modern battery capacities and technologies in the charging system. The major barriers include consumer-range-anxiety or the fear that a vehicle will not have enough energy to travel to its intended destination plus the fact that conventional conductive (plug-in) and more recently, static wireless charging (SWC) has partially alleviated the problem of convenience and safety but does not fundamentally change the operating paradigm that requires vehicle to spend considerable periods of time to receive an energy charge.

Dynamic Wireless Power Transfer (DWPT) or in-motion or on-the-go charging is a paradigm shift in the EV charging ecosystem, also referred to as roadway-embedded infrastructure to a moving vehicle, thus allowing on-the-go or semi-continuous energy replenishment [1]. Therefore, DWPT provides a radical improvement to the fundamental weaknesses of EVs, with the main contribution being the possibility to significantly decrease the on-board battery capacity that is necessary. With its full implementation, EVs can be powered to drive with a smaller, lighter, cheaper battery pack, which means that the EV acquires a theoretically unlimited range on electrified roads, thus alleviating range anxiety and making the EV user experience similar to, or even better than, that of internal combustion engine vehicles [2].

The principles of DWPT are based on the concept of near-field wireless power transfer (WPT) that has been studied for over a century and now achieves the power densities and efficiencies necessary for automotive applications. Recent developments in power electronics, magnetic couplers design and intelligent control architectures, coupled with concerted international standardisation efforts by the Society of Automotive Engineers (SAE) and the International Electrotechnical Commission

R. Agnihotri, *Electrified Road Systems (ERS) for Emerging Asia*, Springer Tracts on Transportation and Traffic 24, https://doi.org/10.1007/978-3-032-16927-3_3

(IEC) to guarantee safety and interoperability, have led to modern systems of DWPT able to transmit tens to hundreds of kilowatts with efficiencies up to 90%.

The chapter provides an in-depth analysis of the concepts of Dynamic Wireless Power Transfer, technologies and applications. It starts by explaining the basics of the physics of magnetic resonance coupling used to enable efficient power transfer across the air gap between the roadway infrastructure and the vehicle. Subsequently, it tears down the design of a typical DWPT system, breaking down the role of each component in the system, starting with the grid connection and all the way to the vehicle battery. The chapter then reviews the new technological advancements of 2020–2025 that can help improve the performance of the system, its scalability, and interoperability. The comprehensive comparison of various applications, i.e., passenger cars, city buses, and long-haul freight trucks, is provided with the background of large-scale international pilot projects. This is followed by a critical analysis of the current technical, economic, and regulatory issues that persist, as well as the mitigation measures that are being developed. The chapter ends by offering a future outlook of the DWPT and how it can be combined with smart grids, renewable energy, and self-driving vehicles and pinpointing several research gaps that will define the future of the technology as it reaches the stage of wide-scale commercialisation.

3.2 Fundamental Principles of DWPT

The ability to conduct a large amount of power to a moving vehicle without a physical touch is based on certain principles of electromagnetism. Although the idea of wireless power can be traced to the work of Nikola Tesla, it was not until the advancements in magnetic resonance coupling that it became possible to use it in high-power automotive applications. This part explains the physics of DWPT, explains the importance of the critical aspect of the quality factor, and gives a comparative insight into the same with respect to static wireless charging.

3.2.1 Electromagnetic Induction and Magnetic Resonance Coupling

Inductive Power Transfer (IPT) is a near-field WPT modality, which is the core technology of DWPT. The main principle of IPT is dictated by the law of Ampere and the law of induction of Faraday. An electric current flowing through a primary (transmitter) coil at varying times creates a varying magnetic field; when a second (receiver) coil is placed in it, the changing magnetic flux through it produces an electromotive force (voltage) which causes a current to flow and power some load or charge a battery [3]. This is very similar to the workings of a standard transformer. However, in a direct inductive coupling system, the power transfer efficiency is very sensitive to magnetic coupling strength between the coils, and so it decays quickly with distance and misalignment [3].

The development of basic induction to magnetic resonance coupling was a turning point in technology, and WPT of automobiles, and consequently, DWPT, became possible. The first inductive systems could only work with millimetre-sized air gaps; automotive systems have much larger gaps of the order of 100–250 mm to allow vehicle ground clearance [1]. Non-resonant inductive coupling is too weak to allow over such a distance.

This drawback can be alleviated by Magnetic Resonance Coupling (MRC), which incorporates capacitors either in series or in parallel with the transmitter and receiver coils, and so forms a pair of LC resonant circuits, or tanks [4]. When the pair of tanks is tuned to the same resonant frequency, they can exchange energy efficiently despite weakly coupled (i.e. large separation) coils [5]. This effect is similar to that of acoustic resonance, in which a tuning fork inductively couples with an adjacent tuning fork of the same natural frequency. Under the MRC system, a transmitter coil induces a strong and oscillating, non-radioactive magnetic field in its close field. This near-field energy is efficiently picked up by the resonant receiver coil, and non-resonant objects in the near field interact only loosely, thus allowing the transfer of kilowatt-level power over distances of several hundred millimetres, resulting in grid-to-load efficiencies to well above 90% [6].

A critical requirement for efficient MRC is precise frequency synchronisation. The resonant frequency (f_r) of an LC circuit is determined by its inductance (L) and capacitance (C), according to the formula:

$$f_r = \frac{1}{2\pi\sqrt{LC}}$$

To achieve optimal power transfer, the operating frequency of the high-frequency inverter on the primary side should be identical to the resonant frequency of both the transmitter and receiver resonant tanks [7]. In electric-vehicle (EV) charging, a nominal frequency of 85 kHz has been standardised in the SAE J2954 specification. This frequency is a trade-off among a number of considerations: it is large enough to allow power transfer efficiency with relatively small magnetic components, and yet small enough to be handled by the current generation of power electronics and to reduce undesirable electromagnetic emissions [8].

3.2.2 Comparative Analysis: Dynamic Versus Static Wireless Charging

Although the principles used in both types of wireless charging are the same, i.e. the principles of WPT, the differences in the way in which they work generate different features, applications to be used and challenges. In comparison to DWPT, in Static

Wireless Charging (SWC), power is transferred to a parked and stationary vehicle as it passes a single charging pad set on the road.

High and more stable efficiency is typical of SWC systems since the vehicle can be perfectly aligned with the charging pad and therefore achieve the highest possible coupling coefficient and minimum air-gap variations. SWC infrastructure is also less complex and requires small, discrete charging units that are installed in special places like residential garages, commercial depots, and parking lots. Nevertheless, although SWC makes it more convenient as it eliminates the use of cables, it does not resolve the problem of range anxiety in any fundamental way because the range of the vehicle is still determined by the amount of energy in its battery between the charging stops.

Instead, DWPT provides the transformative promise of unlimited range with equipped roadways, thus allowing much smaller on-board batteries to be used [9]. This reduction of about 75–90% in vehicle weight, cost, and environmental impact of battery manufacturing can be achieved at the cost of a highly complex technical implementation and high infrastructure investment. DWPT systems also face a moving vehicle with a constantly changing coupling coefficient, resulting in possible power pulsations and system efficacy depending on vehicle velocity and placement in the lane [10]. The infrastructure is much more comprehensive, necessitating the inclusion of long arrays of coils, power electronics and control systems into the public roadways (Table 3.1).

3.3 System Elements and Architecture

A fully operational DWPT system is a complicated combination of power electronics, magnetic parts and control systems broken into two systems: a primary (ground-side) and a secondary (vehicle-side) system. It would be important to learn the role and interaction of each component to understand the possibilities and difficulties of the technology.

3.3.1 Overall System Architecture

The system architecture comprises the principles and designs of the overall system architecture. As illustrated in Fig. 3.1, this process involves multiple stages of power conversion.

- **Primary System (Ground Assembly, GA)**: This is started at the roadside where the utility power supply is tapped. A DC/AC rectifier is a device that transforms alternating current (AC) of a grid into direct current (DC). This DC power is, in turn, fed to a high-frequency power inverter, which produces a high-frequency AC (e.g. 85 kHz) needed to power the transmitter coils. This high-frequency

Table 3.1 Comparative analysis of static versus dynamic wireless charging systems [11]

Feature	Static wireless charging (SWC)	Dynamic wireless charging (DWPT)
Principle of operation	Power transfer to a stationary, parked vehicle	Power transfer to a vehicle in motion
Typical efficiency	Higher and more consistent (90%) due to fixed alignment	Variable (typically 85–92%) and dependent on speed, alignment, and load
Infrastructure	Discrete charging pads in garages, parking lots. Simpler installation	Coils embedded in roadways, complex control systems. High upfront cost
Coil alignment	Fixed and optimized for maximum coupling	Dynamic and variable, posing a significant technical challenge
Battery size impact	No direct impact; requires large batteries for long range	Enables significant battery downsizing (up to 75–90%), reducing vehicle cost and weight
User convenience	High at charging locations (no cables)	Unparalleled; seamless "charge-while-driving" eliminates charging stops
Primary application	Residential, workplace, and public parking	Highways, urban corridors, bus routes, logistics hubs, and ports
Key challenge	Does not solve range anxiety	High infrastructure cost, technical complexity, and standardization

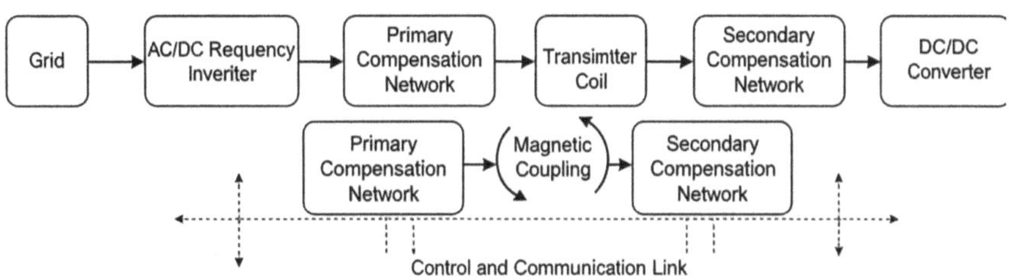

Fig. 3.1 Generic architecture of a dynamic wireless power transfer (DWPT) system

current is conducted through a primary compensation network, and through this, the transmitter coils in the pavement are energised [12].

- **Secondary System (Vehicle Assembly, VA)**: The magnetic field created by the transmitter coil leads to a high-frequency AC being induced in the receiver coil located on the bottom of the EV. This causes current to pass via a secondary compensating network and is then rectified up to DC by an on-board rectifier. Lastly, a DC/DC converter adapts the voltage and current to the needs of the battery of the EV that is controlled by the Battery Management System (BMS) [12].

- **Control and Communication System**: A necessary part of the communication and control logic coordinates the whole process. The presence of a vehicle is detected by vehicle detection sensors (e.g., RFID, GPS), and this can be used to determine the position of the vehicle, which prompts the system to turn on the relevant transmitter coils. The infrastructure and the vehicle are connected through a communication link, in most cases through Vehicle-to-Infrastructure (V2I) protocols, by which the vehicle and infrastructure exchange information about power needs, alignment, and billing, which allows the safe and efficient transfer of power [5].

3.3.2 Roadside Infrastructure: The Transmitter System

The ground system is made up of the power-electronic units, which energise the transmission system and the magnetic structures that are built into the road system. The choice of the architectural make-up of these constituents has far-reaching impacts on the efficiency of the systems, cost, and scalability.

Power Electronics: The major piece of the roadside power-electronics subsystem is a high-frequency inverter, traditionally implemented with advanced power semiconductor switches [13]. To work with high efficiency at the high frequency and power levels of dynamic wireless power transfer (DWPT), modern systems are increasingly using wide-bandgap (WBG) semiconductor power switches like silicon carbide (SiC) and gallium nitride (GaN). These materials provide lower switching losses and higher operating temperatures and higher power density compared to traditional silicon devices, and hence allow inverter designs of smaller size and increased efficiency.

Transmitter Coil Configurations: This is a very important aspect of DWPT infrastructure that determines how the in-road transmitter array should be configured. A key trade-off between infrastructure simplicity and operational efficiency in both the architecture has a set of advantages and disadvantages that are essentially the core trade-off between them, with consequences of considerable importance to electromagnetic interference (EMI), grid impact and economic feasibility.

1. **Long-Track Architecture**: In this design, a single coil of transmitter running continuously is used, potentially covering tens or hundreds of metres long [14]. The major benefit is that it produces a relatively uniform field of magnetic field, and this field can be used to supply a more stable power supply to a moving vehicle. Nevertheless, the method has a high level of inefficiency since the whole coil has to be energised, whether a vehicle is present or not. This continuous energisation results in large power losses and radiating magnetic field that is large and causes a lot of EMI and safety hazards [15].
2. **Segmented-Track Architecture**: A more modern method involves the use of arrays of shorter, discrete transmitter coils, also known as pads, placed along the roadway sequentially, and powered either by their own inverter or by a shared

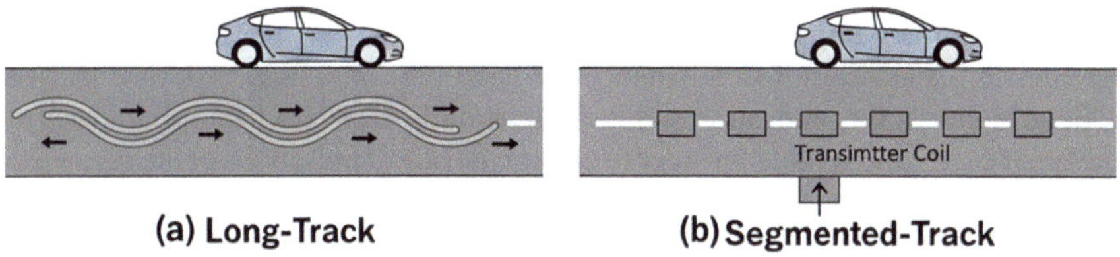

Fig. 3.2 Comparison of **a** long-track and **b** segmented-track transmitter coil topologies

system, switched on only when a vehicle is sensed over it. This localised activation significantly enhances the efficiency of the whole system by reducing the loss of standby power, and also reduces stray electromagnetic fields significantly [16]. The modularity of this design also facilitates the installation, maintenance and scalability. The main disadvantage with it is that it can experience power pulsations as the vehicle crosses over segments where the magnetic coupling reaches zero temporarily [17]. Industry opinion mostly prefers the segmented one on the basis that the inefficiencies of the long-track method, coupled with its safety and EMI concerns, make it virtually unfeasible to deploy the system to large populations.

3. **Coil Design and Materials**: The geometry of the transmitter coils and the materials constituting the transmitter coils determine the performance of the transmitter coils. Several types, such as circular, square and Double-D-D (DD) coils have been developed, each with a trade-off between the coupling factor, misalignment tolerance, and the magnetic field profile [18]. DD and other bipolar pad designs are often favoured because of their ability to focus the magnetic field and their ability to tolerate lateral misalignment. Litz wire is normally used to wind coils and minimise resistance to alternating current at high operating frequencies. They are commonly supported with ferrite cores-materials of good magnetic permeability that direct and focus the magnetic flux toward the receiver, and therefore increase the value of the coupling coefficient and isolate the ground against stray fields. The whole assembly should be well encased and integrated with the pavement construction, with specialised cement-asphalt mortar or magnetisable concrete that would be able to sustain vehicular traffic and exposure to the environment without the magnetic parts being compromised [2] (Fig. 3.2).

3.3.3 Vehicle-Side Components: The Receiver System

The vehicle assembly (VA) will be appointed to extract the magnetic energy from the roadway and feed it to the battery in a secure and controlled way. The nature of its design is limited by the vehicle packaging needs, such as the weight, volume and ground clearance.

Receiver Coil (Pickup) Design: The receiver coil forms the magnetic interface of the vehicle. Its structure has to balance the need to have a large working space to achieve high flux magnetisation and the practical restriction of locating it under the undercarriage of the vehicle. In order to control the power pulsations of the segmented track system, a more sophisticated receiver design has been introduced in the form of multi-coil designs, i.e. a system with two independent coils at the front and the rear of the vehicle. In a more continuous and steady power transfer, an intelligent switching of the coils between the best one to align with the active ground pad is achieved [19]. Multi-pad arrays are also used on the receiver side to achieve maximum energy capture [20]. The receiver coil assembly also implements required shielding. The coil is covered with a layer of ferrite to direct the magnetic flux coming into it, and a conductive plate, usually aluminium, is placed over the ferrite to cover the chassis and cabin of the vehicle against any remaining magnetic fields by creating opposing eddy currents [21].

Power Electronics: The vehicle side power electronics condition the energy obtained to charge batteries. The alternating current of the receiver coil is converted to direct current by a high-frequency rectifier (typically an active rectifier that uses controlled switches to achieve higher efficiency) followed by a DC/DC converter (e.g., a buck converter or boost converter) that does a critical control task: it regulates the output voltage and current such that charging to the battery management system (BMS) charging profile is achieved, to ensure safe and optimum charging in varying states of charge and battery temperatures [5]. A recent development in VA design is to combine these functions with the existing on-board charge. This approach leads to a reduction of duplication, cost and weight, and complexity of the components by utilising the preexisting rectifier and DC/DC converter steps of the OBC, with the VA basically being a wireless front-end (Fig. 3.3).

3.3.4 Compensation Topologies

Compensation topologies are those that are made of networks of capacitors (occasionally with inductors) that are connected on either side to the transmitter and receiver coils. They play multiple essential roles in the workings of a resonant wireless power transfer (WPT) system: they compensate the large self-inductance of the coils at the operating frequency, therefore, permitting resonance; they minimise the reactive power required to drive the inverter; and they can be designed to provide desirable output properties, such as constant voltage or constant current regardless of load or coupling variations.

These four basic topologies of compensation are labelled with the series (S) or parallel (P) connection of the capacitor on the primary and secondary sides: Series (SS), Series Parallel (SP), Parallel Series (PS), and Parallel (PP).

- **Series-Series (SS)**: This is another common type of connection that is used in charging electric vehicles (EVs). The major benefit of it is that the capacitance

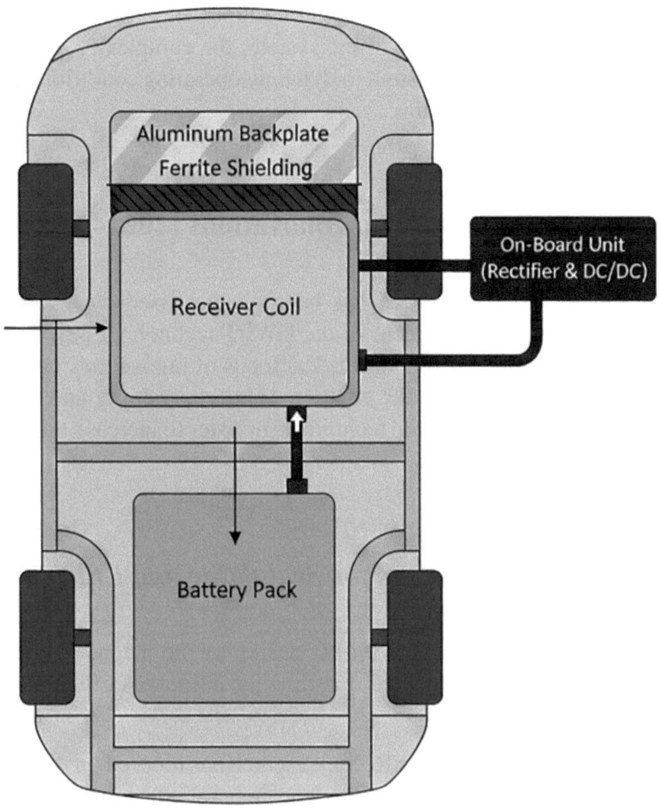

Fig. 3.3 Key components of a vehicle-side receiver assembly

required does not depend on the coupling coefficient or the load, and the contribution to the current induced on the secondary side is not much dependent on the load, making it particularly well-suited to battery charging [8].
- **Series–Parallel (SP)**: This circuit design provides a fixed-voltage output characteristic, which may be desirable to put a voltage bus directly on the load, though the capacitance on the primary side varies with load and coupling factor.

In order to address the unique issues of dynamic wireless power transfer (DWPT), including the broad range of coupling and load, more advanced, higher-order topologies have been developed.

- **LCC Topology (Inductor-Capacitor-Capacitor)**: The LCC Topology is commonly used in the LCC form (LCC on the primary side, series on the secondary side) or the Double-Sided LCC, which gives the constant-current output on the primary side not influenced by the coupling factor. It therefore demonstrates an extraordinarily strong ability against misalignment and can be used to propel a transmitter coil array in a DWPT system [9].

The choice of a compensation topology is a highly important design choice that has a direct effect on the efficiency of the system, the complexity of control in the system and the resilience of the system to dynamic operating conditions experienced in the process of on-the-go charging.

3.4 Recent Developments and Innovations (2020–2025)

The period between 2020 and 2025 has been characterised by a lot of research and development on the development of the DWPT technology beyond pilot-scale demonstrations to commercial feasibility. The focus of innovations has been on the improvement of core ingredients, the creation of interoperability and the utilisation of superior materials and calculation techniques in order to increase the performance of the system.

3.4.1 Advanced Coil Designs for Scalability and Efficiency

Innovation is primarily concerned with the design of the magnetic couplers, both on the road and in the vehicle, since this directly influences efficiency, cost, and scalability.

- **Modular and Segmented Coils**: Recent research has focused on modular designs of transmitters in order to enable large-scale implementation. These systems consist of standard, pre-fabricated coil segments which can easily be installed, connected and extended so as to lower the complexity and expense of electrifying long sections of roadways [14]. This modular strategy is a requirement to realise the economies of scale that are needed to be adopted in large numbers.
- **Self-Decoupled and Integrated Coils**: The major technical hurdle in dense, segmented arrays of transmitters is the electromagnetic cross-coupling that takes place between adjacent coils. This reciprocal inductance has the potential to destabilise power inverters and make it difficult to control. The major innovation to solve this is by creating self-decoupled coils that add auxiliary or compensation windings linked in series with the main transmitter coil. The auxiliary windings are designed in a geometrical distribution to create a magnetic flux that exactly cancels the flux of neighbouring coils, thus separating each segment. Such designs have been experimentally proven to reduce cross-coupling by up to 95% which is a vital breakthrough to allow robust and scalable multi-transmitter systems.
- **Multi-Coil and Reconfigurable Receivers**: To reduce the natural power variation that is realised when a vehicle passes the gaps between transmitter segments, state-of-the-art receiver designs are being developed. A potential solution is the use of two or more independent receiver coils, e.g. one at the front and one at the back of the vehicle. An intelligent control system intelligently chooses between using coils

with the optimum alignment with the active ground pad at any particular point in time. The method is known to maintain an almost constant coupling factor, hence stabilising the power and efficiency transferred. Simulation experiments have demonstrated that a two-coil system can also minimise the fluctuation of powers to only 2.8% as compared to the severe power holes experienced with a conventional single-coil receiver [19].

3.4.2 Interoperability and Standardisation

To become ubiquitous as a utility in society, DWPT should work well with different brands of vehicles and infrastructure providers. This would require combined efforts in the areas of technological development and international standardisation. These areas have been developed in parallel, with solutions to each being demonstrable, standards formalising them into a commonplace architecture, creating a competitive, multi-vendor marketplace.

- **The Challenge of Interoperability**: One of the biggest obstacles to market penetration would be the lack of a standard that is globally accepted as the magnetic interface. Various manufacturers have formulated some proprietary coil topologies (e.g. circular unipolar, DD bipolar), and a vehicle using one type of receiver may not be able to be charged effectively, or at all, by another transmitter.
- **Technological Solutions for Cross-Compatibility**: Studies are guided to produce universally compatible parts. This also entails new coil recipient designs that are built to achieve intrinsic interoperability. As an example, a Vertical Overlapping **Decoupling (VOD)** coil made of double coils can be overlaid with different paths of magnetic flux to provide high interoperability with DD, square, and circular transmitters, with an efficiency greater than [22]. The other approach involves the use of reconfigurable systems, in which the power electronics of the vehicle dynamically select between various coil segments or a varying compensation network to provide the vehicle with an electrical fit to the observed infrastructure, thus keeping all performances optimum, of the transmitter type.
- **Development of International Standards**: In response to the urgency of having an international framework, international standards organisations are actively coming up with specifications of wireless power transfer (WPT). Although the emphasis has been laid on the concept of static charging, the standards are currently being extended to suit the special needs of dynamic charging. Table 3.2. A summary of some of the fundamental standards that govern WPT among electric vehicles (EVs) is given. SAE J2954 gives the technical specifications of permanently parked wireless charging of light-duty vehicles and sets up principles of interoperability, safety, and performance up to 11 kW, with the forthcoming SAE J2954/3 specifying specifically dynamic use of light-duty vehicles. Simultaneously, the IEC 61980 series offers a detailed standard of magnetic-field WPTs and the future IEC PAS 61980 5 and 6 address the issue of interoperability, safety, and communication protocols of dynamic WPT. 63 ISO 19363 supplements them

Table 3.2 Key international standards governing WPT for electric vehicles [24]

Standard	Issuing body	Scope and relevance to DWPT	Status (as of 2024/2025)
SAE J2954	SAE International	Establishes specifications for interoperability, safety, and performance for *stationary* WPT up to 11 kW for light-duty vehicles. Forms the foundation for dynamic standards	Published (2024 update)
SAE J2954/2	SAE International	Technical information report (TIR) for *stationary and dynamic* WPT for heavy-duty vehicles, with power levels up to 500 kW	TIR published (2022)
SAE J2954/3	SAE International	Planned standard specifically for *dynamic* WPT applications for light-duty vehicles	In development
IEC 61980 series	IEC	Covers general requirements, communication, and specific requirements for MF-WPT systems. Harmonized with ISO standards	Parts 1, 2, 3 published
IEC PAS 61980-5 and 6	IEC	Publicly available specifications focusing specifically on interoperability, safety, communication, and activities for *dynamic* MF-WPT systems	PAS 61980-6 scheduled for 2025
ISO 19363	ISO	Defines safety and interoperability requirements for the *on-board vehicle equipment* for MF-WPT, intended to work with IEC 61980-compliant infrastructure	Published (2020/2021). Does not cover in-motion charging in the current edition

with the specification of the on-board vehicle equipment to ensure that vehicles are equipped to operate with IEC systems [23].

3.4.3 *Advanced Materials and AI-Based Optimisation*

It is expected that the breakthroughs in the field of materials science and the usage of artificial intelligence (AI) will lead to future performance gains in dynamic WPT (DWPT).

- **Advanced Materials**: The effectiveness of the magnetic coupling is highly dependent on the properties of the materials that are to be used. The current studies explore more developed ferrite alloys that have a higher magnetic permeability and lower losses with high frequencies and can therefore better focus the magnetic flux and increase the Q factor of the coils. Also, new pavement materials, including magnetisable concrete, where ferrite particles are included in the cement matrix, are being considered as a multi-purpose structural road layer and magnetic core, which could potentially be cheaper and easier to install. In the future, a possible direction of exploiting the possibility of manipulating magnetic fields with even greater precision is through metamaterials, which are artificially realised structures possessing electromagnetic behaviour not found in natural materials, including a negative refractive index. In the early research stages of WPT, it is possible that once the metamaterials are fully developed, they can be used to form a magnetic field lens, bending it into the receiver coil or even highly effective shielding solutions that are fully lightweight.

- **AI-Based Optimisation**: AI and machine learning (ML) are the best solutions to optimisation because of the complexity and dynamic nature of DWPT systems. The AI-controlled system will be able to accept real-time information on vehicles and the grid to make smart decisions that will boost overall performance. Potential applications include:

  Intelligent Coil Activation: AI can estimate vehicle paths and velocities in order to open and close transmitter coils at the right time to maximise the amount of energy transfer and minimise energy wastage [25].
  Dynamic Load Balancing: AI can estimate vehicle paths and velocities in order to open and close transmitter coils at the right time to maximise the amount of energy transfer and minimise energy wastage [26].
  Adaptive Power Control: ML models can be trained to figure out the correlation between misalignment speed and power-transfer efficiency in real time and control the inverter output dynamically to keep the charging optimality in response to the changing driving conditions.

3.5 Applications and Deployment Scenarios

To gain a clear insight into the transformational power of DWPT, it is more appropriate to consider its application to particular transportation sectors. Both of the use cases capitalise on the three primary advantages of the technology: range extension, reduction of the battery size, and automation of operations to address different problems. Live pilot programs in various countries around the world are now proving the performance and feasibility of DWPT in these various situations.

3.5.1 Passenger EVs: Range Extension and Battery Downsizing

In the case of the passenger-car market, the key strength of DWPT will be its ability to ensure a solution to the range anxiety as a significant barrier to EV usage. By allowing vehicles to charge as they pass along major highways and urban arteries, DWPT will essentially guarantee the delivery of an unlimited driving range, making

EV usage as comfortable as that of a conventional vehicle. This is made possible by direct facilitation of a paradigm shift in vehicle design: the shift towards much smaller on-board battery packs. The battery is part of the EV that can be attributed to a significant percentage of its cost, weight, and environmental footprint. Once a vehicle can habitually be charged during motion, it does not need a huge battery that is designed to last 300–400 miles when charged. Rather, it will be able to work with a significantly reduced battery pack that will be enough to cover the distance between electrified corridors and the last few miles to a destination. Research and analyses indicate that extensive DWPT infrastructure would provide the possibility to considerably reduce battery capacity of up to 75–90%, not only to make EVs cheaper and more reachable, but also to make them less heavy and more energy-efficient, which would create a cycle of sustainability.

3.5.2 Urban Public Transit: Continuous Charging Corridors

DWPT is highly applicable when it comes to urban mass transit, especially buses that use electricity. These cars are usually on regular, predetermined routes, and thus utilise the inbuilt charging infrastructure quite heavily, thereby enhancing the payoff of a large capital outlay. Operators can use a technique of continuous opportunity charging to implement by installing DWPT systems along portions of a bus route. This enables buses to charge their batteries during the day at the service, eliminating the need to go back to a central depot to charge batteries after a long period. This operational model has several major benefits: buses can operate effectively 24/7, fleet size is minimised (which serves routes), and the physical area taken up by depot charging infrastructure is reduced.

- **Case Study: KAIST OLEV Project (Gumi, South Korea)**: The Korean Advanced Institute of Science and Technology (KAIST) came up with one of the first and most prominent examples of the use of the concept of dynamic wireless power transfer (DWPT) in a public transport setting with the Online Electric Vehicle (OLEV) project. In 2013, the city of Gumi fitted buses that are equipped with OLEV on a 24 km inner-city route. This system was designed to conduct 100 kW of power and had a reported efficiency of 85% over a 17 cm air gap [27]. The technology, which was known as the Shaped Magnetic Field in Resonance (SMFIR), used segmented power tracks that occupied only 5–15% of the overall route, and thus proved the possibility of intermittent charging [28]. Even though the project was able to maintain the public transportation for several years, and it was even a technological breakthrough, it also highlighted the problems in the long term. Some of the earlier bus lines would later be abandoned as the expenses of maintaining the old in-road facilities were costly, thus emphasising the relevance of durability and cost of lifetime in DWPT deployments [27].

3.5.3 Long-Haul Freight and Logistics: Reducing Downtime

The electrification of the heavy, long-haul freight trucks is a daunting task because these vehicles consume a lot of energy, and therefore, would require prohibitively large, heavy, and expensive battery packs. Importantly, downtime cost is an enormous economic cost in the logistics sector, where the most important factor is the utilisation of assets [29]. DWPT provides a good alternative because it will allow receiving power continuously as vehicles pass along major freight routes. The trucks may have smaller and easier-to-handle battery packs and be fed the necessary propulsion energy by the road as they travel at highway speeds [30]. This plan would significantly lower the duration of idle time spent at the high-power charging stations, which would lower the expensive idle time and allow the trucks to meet strict delivery times. Furthermore, having a smaller battery pack will minimise the weight of the vehicle, which will consequently allow a higher payload capacity, which is an undeniable economic benefit to freight operators.

- **Case Study: INDOT/Purdue University Project (Indiana, USA)**: INDOT and Purdue University are the pioneers in the pilot project after identifying the possibility of heavy-duty trucking. In 2024, a quarter-mile testing facility on a public highway (US231/52) was built in West Lafayette. The project aims at testing and confirming a patent-pending DWPT system that can transfer more than 200 kW of power to an electric truck with heavy loads running at up to 65 mph [30] of highway speeds. This project is one of the first projects in the world that addresses the high-power needs directly of the Class 8 trucks, and it will provide important information about the performance, durability and the interaction of the technology with the pavement structures [31].

3.5.4 Other Applications: Port Automation and Autonomous Vehicles

These advantages of automated, contactless charging are extrapolated to other specialised industrial and future mobility.

- **Automated Guided Vehicles (AGVs) and Drayage Operations**: AGVs are becoming common in the container transport of controlled areas like seaports and logistics centres. A continuous power supply for these vehicles can be made available through DWPT, which allows the vehicles to operate 24 h, without any manual charging or change of batteries. This increases the efficiency of operation, lowers labour expenses and forms a major enabler to the provision of fully automated port operations [11].
- **Autonomous Vehicles (AVs)**: DWPT is an ideal technological collaborator of a complete AV transportation ecosystem. To be self-reliant in motor ability, AVs need to regulate themselves with regard to their energy requirements. DWPT provides the other half of the puzzle, a complete automated energy replenishment

system. When its power had depleted, an AV might drive on an electrified lane successfully, recharge itself on the way, and keep on going without human control.

3.5.5 *Global Pilot Projects and Performance Metrics*

The practicability of DWPT is being proven by the increasing number of pilot projects all over the world. These tests provide priceless data about the performance, reliability and real-world operating issues of the system. Table 3.3. represents the summary of key metrics of various major international projects. The Arena del Futuro project in Italy, which consisted of a 1.05 km track on the A35 Brebemi motorway, showed that a passenger EV would be capable of speeding on the highway using a net-zero amount of battery power, with the charging efficiency of a DC fast charger [32]. Electreon, a technology provider, is currently installing its technology on several roads in several countries, such as a 1 km e-bus road in Balingen, Germany, and the first public dynamic charging road in the United States, located on a 1 mile stretch in Detroit, Michigan [33]. These, and the heavy-duty truck focus of the INDOT/ Purdue testbed, together with the pioneering work of the KAIST OLEV project, are all demonstrations of the versatile nature of the technology and its development towards commercial viability.

3.6 Challenges and Mitigation Strategies

Even with its paradigm shift potential, the ubiquitous implementation of DWPT is accompanied by serious challenges in technical, economic and regulatory fields. These issues are mutually reinforcing, and they make up a trilemma of efficiency, cost and safety; innovation in one field often puts a constraint on the others, and the overall, system-level perspective on innovation is essential.

3.6.1 *Technical Challenges*

The main technical issue of DWPT is to provide high amounts of power to a high-speed vehicle in variable conditions with high efficiency and reliability.

- **Efficiency and Power Transfer**: It is hard to sustain a grid-to-battery efficiency of more than 90% in a dynamic environment. There is a loss of power in each conversion step, both in the power electronics, in the magnetic coupling and in the coils themselves. At highway speeds, transmitting high power levels (e.g. 200 kW in trucks) on the highway will require the car to take only a fraction of a second per transmitter segment, which implies it will be necessary to transfer the energy with extremely high speeds and efficiency [10].

Table 3.3 Performance metrics of major international DWPT pilot projects [1]

Project name (location)	Key partners	Track length	Power level	Vehicle speed	Key findings and status
Arena del Futuro (Italy)	Stellantis, A35 Brebemi, ElectReon	1.05 km	1 MW (system)	Highway speeds	Demonstrated charge-sustaining operation for passenger car; efficiency comparable to DC fast charging. Live
INDOT/Purdue Testbed (USA)	INDOT, Purdue, ASPIRE	0.25 mile (~0.4 km)	>200 kW (target)	65 mph (target)	Focus on heavy-duty trucks; construction began in 2024. In progress
ELINA Project (Germany)	ElectReon, EnBW, KIT	1 km	Not specified	City speeds	Focus on public e-bus; validation of planning tools. Live
Michigan Pilot (USA)	MDOT, ElectReon, Ford	1 mile (~1.6 km)	Not specified	City speeds	First public dynamic charging road in the U.S.; testing with shuttles and delivery vans. Live
KAIST OLEV (South Korea)	KAIST, Gumi City	24 km (route)	100 kW	City speeds	Pioneering e-bus deployment; operated for several years but faced challenges with infrastructure aging. Some routes shut down

- **Misalignment Tolerance**: A moving vehicle is by definition laterally and longitudinally misaligned with the centre of the transmitter coils, which is not the case with the stationary charging. Even in the deviations as small as the one depicted in Fig. 3.4, a sudden decrease in the magnetic coupling coefficient is likely to cause drastic changes in power swings and lower transfer efficiency [10]. This inconsistency creates a significant barrier to the ability to have consistent and reliable charging.
- **Durability and Maintenance**: In-road components of the dynamic wireless power transfer (DWPT) system require very high robustness. They must survive in a state of incessant mechanical strains due to heavy traffic, large-scale temperature changes, and exposure to water, sodium chloride and other environmental elements throughout a service life of multi-decadal. The robustness of the transmission coils, power electronics thereof and their connection to the vehicle pavement is a very dangerous engineering challenge [31]. Furthermore, the logistics and economic challenges of carrying out maintenance or repair interventions on the subsurface elements that lie beneath a highway lane are significant, requiring the closure of the lane and the use of specialised equipment.

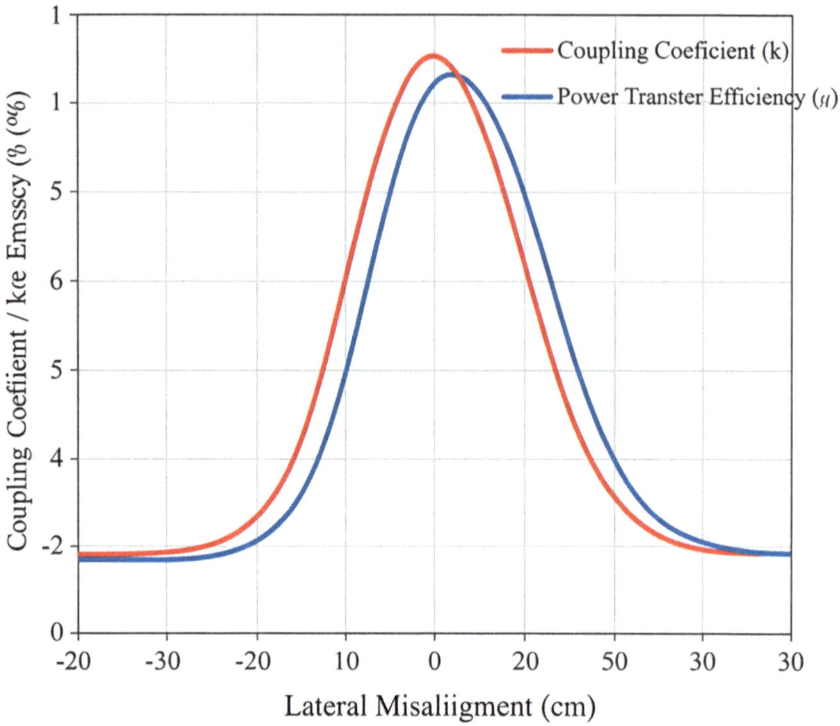

Fig. 3.4 Impact of lateral misalignment on coupling coefficient and power transfer efficiency

- **Mitigation Strategies**: Technical solutions to such challenges are being developed currently by researchers. To reduce the effects of misalignment, advanced control schemes like field-oriented control are used that allow the currents in the multi-coil transmitters to be electronically adjusted, thus steering the magnetic field in real-time towards the receiver coil. Multi-coil receiver arrays on the car side can maintain a more constant connection when the car goes through different transmitter arrays. Adaptive impedance-matching networks also dynamically optimise the tuning of the system to counter the changes in coupling to maintain optimal power transfer, e.g. by using high-strength, thermally-conductive concrete.

3.6.2 Economic Viability

The most overwhelming obstacle to the popular use of DWPT is its high cost.

- **High Infrastructure and Installation Costs**: The amount of capital to develop and install kilometres of transmitter coils, power electronics and grid interconnections into maintenance roads is high. These costs make DWPT one of the most expensive electric-vehicle charging systems per mile, which is a big burden on both government and personal investments.

- **Return on Investment (ROI)**: It is crucial to create a viable business model that will have a clear entry point into ROI. The financial viability is dependent on a complicated interrelation of factors, such as the original cost of the installation, operation and maintenance expenses, tariffs, electricity and the amount of money that will be collected as a result of the user. One of the most important variables is the rate of utilisation; a DWPT lane that is highly utilised by freight trucks or even by public buses will generate more revenue faster than one that is hardly used by passenger cars [34].
- **Mitigation Strategies**: Deployment strategies will tend to be gradual and focused in order to increase the economic viability. The first projects will focus on high-value, high-utilisation situations where the economic values are most pronounced, i.e. dedicated bus lanes, freight corridors linking key logistics centres to each other, and operational areas in ports and airports. Solutions based on financial models, such as Public–Private Partnerships (PPPs), are also under consideration to share the cost and risk between the government and the industry. Moreover, new business models, including the so-called Charging as a Service (CaaS), where a third-party company collectively owns and manages the infrastructure and provides charging as a service to fleet operators: subscription or pay-per-use, could allow overcoming the barrier of large initial capital outlay [35].

3.6.3 *Safety and Regulatory Compliance*

The safety of both the population and vehicle occupants, as well as the electrical grid, is of the top priority and is controlled by powerful regulatory norms.

- **Electromagnetic Field (EMF) Exposure**: Intense magnetic fields are produced by the high-power DWPT systems. One of the basic safety considerations is that non-ionising radiations should not exceed the prescribed exposure limits as set by international unions like the International Commission on Non-ionising Radiation Protection (ICNIRP) and the IEEE in terms of stray or leakage fields outside the immediate charging zone. This confidence is essential to safeguard people in vehicles, pedestrians, as well as those who carry delicate medical equipment such as pacemakers [30].
- **Electromagnetic Interference (EMI)**: The high-frequency fields may cause interference in both the electronic components of the vehicle (e.g., control units, sensors, and infotainment) and other radio and communication facilities. Limiting the effects of such interference is one of the aspects of electromagnetic compatibility (EMC) compliance [36].
- **Thermal Management**: Transmissive power always heats up as a result of resistive loss in the coils and switching losses in the power electronics. Proper cooling of the heat of the in-road transmitters and the vehicle mount receivers is critical

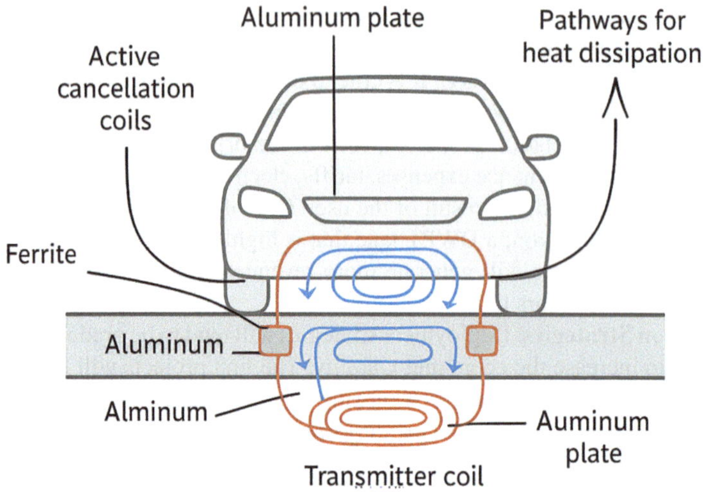

Fig. 3.5 EMF shielding and thermal management strategies for DWPT systems

to avoid overheating, which may lead to poor performance of the components, shorten their life and in worst-case scenarios, fire may occur.

- **Mitigation Strategies**: There is a multi-layered scheme of EMF, EMI and thermal issues, which is shown in Fig. 3.5. The first line of defence is passive shielding, where magnetic flux and conductive plates (usually aluminium) are used to confine leakage fields through the use of ferrite materials in the magnetic flux guiding field, and heat sinks in the thermal management design, or thermally conducting materials in the in-road example [21]. This is often supplemented by active shielding, where the leakage fields are actively cancelled by the use of additional cancellation coils in the magnetic flux guiding field.

3.7 Future Outlook

The further evolution of the DWPT technology will be marked by greater integration into the broader transportation and energy ecosystems as it matures. The long-term vision does not just consider charging vehicles, as it will make electrified roadways an active, smart, and supplementary part of the future smart cities and clean energy networks.

3.7.1 Integration with Smart Grids and Renewable Energy

The mass installation of dynamic wireless power transfer (DWPT) systems will also present a major and intermittent additional load on the electrical system. This demand should be managed successfully to ensure grid stability. This is a challenge and at the same time, a chance to integrate synergistically with smart grid technologies. The smart grid management systems can liaise with the DWPT infrastructure to plan and control charging loads on the fly. An example of this would be to give more power to the areas of the grid that are electrified to roads and reduce during times of peak grid demand, thus helping to balance the grid and make the most of the available clean energy [25].

Further into the future, the emergence of bidirectional power transfer, or Vehicle-to-Grid (V2G) operations to establish DWPT systems, will be a vital research direction [9]. In a bidirectional system, a national grid-scale battery storage of electric vehicles (EVs) passing highways could be used as a distributed energy resource. There would be vehicles that would absorb excess grid energy at times when there is high generation and release parts back to the grid to offer ancillary services to the grid, like regulating frequency or alleviating peak demand. In this hypothetical situation, the electrified highway network is no longer a passive consumer of energy but a vast storage system of energy that is active in the energy market and contributes to the stability and sustainability of the power grid.

3.7.2 Synergy with Autonomous Vehicles and V2X Communication

DWPT and autonomous vehicles (AVs) represent some of the most complementary technologies, which, when integrated, can allow the development of a fully automated transportation ecosystem. AVs should be in a position to have control over their energy logistics in order to achieve their potential of functioning without a human operator. The missing connection that DWPT provides is a smooth, automated way of energy replenishment that does not require manual connection via cable or manual navigation to a charging station. A self-driving electric truck on a long-haul trip, such as one, would be coded to use electrified lanes to keep its charge, thus allowing it to operate continuously without making unscheduled stops or human-charging.

Vehicle-to-Everything (V2X) communication will be used to coordinate this. With protocols like Dedicated Short-Range Communications (DSRC) or Cellular V2X (C-V2X), cars can constantly communicate with the road infrastructure and provide their real-time energy needs, state of charge, and destination. The infrastructure, in its turn, can dynamically price charging services, control the coil actuation to maximise power transmission to the vehicle in accordance with its speed, and even provide routing advice to guide the vehicle in the most energy-efficient route. This data exchange is essential to the successful, efficient and cost-effective operation of a large-scale DWPT network.

3.7.3 *Roadmap for Global Scaling Towards 2040*

The trend of the widespread introduction of DWPT to the world is bound to be a progressive, planned undertaking in the coming twenty years. The expenses and technicalities involved in implementing the technology are high, and this means that a staged implementation is required, which would put more focus on applications that will give the best and most immediate payoff.

- **Phase 1 (Present–2030): Targeted Deployments**. The first phase is concentrated in the controlled high-utilisation settings, including DWPT rollout on special routes of urban bus fleets, in large logistics centres and seaports to supply automated cars, and major, high-traffic freight routes between major economic centres. These first markets are considered as beachhead markets, where the technology can be allowed to evolve, gain economies of scale in production, and demonstrate the economic and operational advantages in actual operation conditions [37].
- **Phase 2 (2030–2040): Corridor Expansion and Network Effects**. With prices falling and the harmonisation of the standards across the border, the deployment will expand to cover larger parts of large highway systems. This growth will create so-called network effects, where the value of a vehicle equipped with DWPT will increase as the number of electrified miles grows. This phase will most probably be spearheaded by public–private partnerships and favourable government policies that are geared towards decarbonising long-haul freight and transportation.
- **Phase 3 (Beyond 2040): Ubiquitous Integration**. DWPT may eventually become part of the urban and highway infrastructure across the entire globe in the long term. With the full integration of the technology with autonomous vehicles and smart grids, the electrified road will become an active, distributed energy network, with an ultimate transformation of the symbiotic relationships between transportation and energy.

This roadmap is conditional on the ongoing technological innovation, creating clear and harmonised international standards, building sustainable business models, and applying supportive government policies.

3.8 Conclusion

Dynamic Wireless Power Transfer (DWPT) is a groundbreaking technology in the field of electrification in transportation, which can offer a valid substitute to the drawbacks of battery-powered electric vehicles as it is. DWPT can help eradicate range anxiety, help to realise a radical decrease in battery size and cost, and create a smooth and automated charging process by allowing energy to be transferred to moving vehicles. The concept of the electric road has become a technical reality due to the principles of magnetic resonance coupling, sharpened by several decades of studies, which allow the free flow of high power rates through the air gap between roadway and vehicle and through the large distance between them.

This chapter offers a synthesis of DWPT, in both its principles of physics and its applications in the real world. System architecture analysis demonstrates that there is a complicated interaction between the high-frequency power electronics, high-technology magnetic designs, and high-technology control systems. The recent advances, such as modular and self-decoupled coils, reconfigurable receivers and work on international standardisation, are still bringing the technology nearer to commercial feasibility. The DWPT's ability is tested in piloting projects around the world with a variety of vehicles (including passenger cars, urban buses, and heavy-duty trucks), producing encouraging results in terms of power transfer efficiency and operational advantages.

However, the application of DWPT is still faced with a lot of challenges. Limitations in technology continue to ensure high efficiency in dynamic environments that are high-speed and misaligned, and one of the major subjects of modern study. Its heavy capital expenditure on the installation of infrastructure poses an intimidating economic impediment that requires novel business approaches and targeted rollouts. In addition, the security of the populace in terms of strict control over electromagnetic fields and guarantees the sustainability of in-road materials in the long term are absolute necessities that require solid engineering principles and clear regulations.

The further development of DWPT is also inseparably connected to the active progress of smart grids and autonomous vehicles. The combination of these technologies will result in an ecosystem of transport that will be not only sustainable and efficient but also intelligent and full of automation. The electric road can be developed further than a surface on which to travel, into a dynamic energy system that proactively sustains the stability of power grids and opens up a new mobile age.

This vision requires further and concentrated research to realise it. The major gaps in the research that should be addressed are:

- Advanced Materials: New low-cost, high-performance materials based on magnetism and pavement-integrated materials are needed, which will improve efficiency, durability, and lower the infrastructure cost.
- Intelligent Control Systems: The concept of AI-based, network-level control of power flow and grid impact, as well as the optimisation of the efficiency of large-scale DWPT systems under real traffic situations, should be further investigated.
- Interoperability and Standardisation: Although the international standards of dynamic charging among all types of vehicles are slowly being harmonised, it is vital to accelerate this process in order to create a competitive and open market and to guarantee a tone-free user experience.

- Techno-Economic and Lifecycle Analysis: In-depth research on business models must be conducted to understand long-term economic feasibility and environmental sustainability, and to have clear data on the matters to inform the policy of the people and the investments.

Finally, Dynamic Wireless Power Transfer represents a technology of huge potential. Despite the fact that the challenges are only too significant, the continued improvement in research and development, as well as standardisation, indicates that DWPT will be able to become a key to a more sustainable, efficient, and convenient future of electric mobility.

References

1. MDPI. Dynamic wireless power transfer charging infrastructure for future EVs: from experimental track to real circulated roads demonstrations. https://www.mdpi.com/2032-6653/10/4/84. Accessed 4 Oct 2025
2. ResearchGate. Design and evaluation of a wireless power transfer system with road-embedded transmitter coils for dynamic charging of electric vehicles. https://www.researchgate.net/publication/325291439_Design_and_Evaluation_of_a_Wireless_Power_Transfer_System_with_Road_Embedded_Transmitter_Coils_for_Dynamic_Charging_of_Electric_Vehicles. Accessed 4 Oct 2025
3. ResearchGate. Static and dynamic wireless charging of electric vehicles using inductive coupling. https://www.researchgate.net/publication/372535378_Static_and_dynamic_wireless_charging_of_electric_vehicles_using_inductive_coupling. Accessed 4 Oct 2025
4. Infineon Technologies. White paper: resonant wireless power transfer. https://www.mouser.com/pdfdocs/Infineon-Whitepaper_WirelessCharging_Wireless_Resonant_Power_Transfer-WP-v01_00-EN-3.pdf. Accessed 4 Oct 2025
5. Scientific Research Publishing. Designing an effective method for automatic electric vehicle charging stations in a static environment. https://www.scirp.org/journal/paperinformation?paperid=130712. Accessed 4 Oct 2025
6. IJRASET. Review of static and dynamic wireless electric vehicle charging system. https://www.ijraset.com/best-journal/static-and-dynamic-wireless-electric-vehicle-charging-system. Accessed 4 Oct 2025
7. IJOER. Concepts, quality factor and efficiency of wireless power transfer systems. https://ijoer.com/assets/articles_menuscripts/file/IJOER-APR-2015-1.pdf. Accessed 4 Oct 2025
8. Wiley Online Library. Recent advances in wireless power transfer for electric vehicle charging: roadway dynamic charging and future trends. https://doi.org/10.1002/est2.278. Accessed 4 Oct 2025
9. IJRPR. Review on dynamic wireless charging of electric vehicle. https://ijrpr.com/uploads/V5ISSUE1/IJRPR21661.pdf. Accessed 4 Oct 2025
10. ResearchGate. A comparative analysis of static and dynamic wireless charging systems for electric vehicles. https://www.researchgate.net/publication/384139876_A_Comparative_Analysis_of_Static_and_Dynamic_Wireless_Charging_Systems_for_Electric_Vehicles. Accessed 4 Oct 2025
11. E3S Web of Conferences. Electric vehicle charge dynamic wireless power transfer. https://www.e3s-conferences.org/articles/e3sconf/pdf/2025/19/e3sconf_icsget2025_01007.pdf. Accessed 4 Oct 2025
12. STM Journals. A review of dynamic wireless charging system for electric vehicles and IoT technology. https://engineeringjournals.stmjournals.in/index.php/TEE/article/view/7465. Accessed 4 Oct 2025

13. IEEE Xplore. Resonant inductive coupling for electric vehicle charging systems—a review of challenges and solutions. https://ieeexplore.ieee.org/document/10067594. Accessed 4 Oct 2025
14. SpringerLink. Design optimization of wireless power transfer coils for electric vehicles. https://doi.org/10.1007/s12647-021-00514-1. Accessed 4 Oct 2025
15. I.R.I.S. Electromagnetic interference in cardiac implantable electronic devices due to dynamic wireless power systems for electric vehicles. https://iris.uniroma1.it/retrieve/6c192afe-de3b-4c24-9e2f-20c3d8b83d25/Campi_Electromagnetic%20Interference_2023.pdf. Accessed 4 Oct 2025
16. ResearchGate. Dynamic wireless power transfer for electric vehicles—technology and infrastructure. https://www.researchgate.net/publication/383847940_Dynamic_Wireless_Power_Transfer_for_Electric_Vehicles_-_Technology_and_Infrastructure. Accessed 4 Oct 2025
17. Elsevier. Inductive power transfer for urban mobility: infrastructure planning for smart cities. https://www.sciencedirect.com/science/article/pii/S221067072100042X. Accessed 4 Oct 2025
18. MDPI. Dynamic wireless charging for electric vehicles—a comprehensive overview. https://www.mdpi.com/1996-1073/15/4/1480. Accessed 4 Oct 2025
19. Elsevier. Dynamic wireless charging infrastructure for electric vehicles—cost optimization and implementation strategies. https://www.sciencedirect.com/science/article/pii/S0306261924001479. Accessed 4 Oct 2025
20. MDPI. Conductive electric road localization and related vehicle power control. https://www.mdpi.com/2032-6653/13/1/22. Accessed 4 Oct 2025
21. MDPI. Precise analysis on mutual inductance variation in dynamic wireless charging of electric vehicle. https://www.mdpi.com/1996-1073/11/3/624. Accessed 4 Oct 2025
22. Volvo Group (2024) Dynamic charging for heavy vehicles: technology roadmap. https://www.volvogroup.com/. Accessed 4 Oct 2025
23. Electreon. Gotland, Sweden electric road case study. https://electreon.com/projects/gotland. Accessed 4 Oct 2025
24. Nordic Energy Research. Electric road systems. https://www.nordicenergy.org/wp-content/uploads/2019/10/Electric-roads_web.pdf. Accessed 4 Oct 2025
25. Electreon. E|MPOWER—Autobahn. https://electreon.com/projects/empower-autobahn. Accessed 4 Oct 2025
26. Smartroad Gotland. Home. https://www.smartroadgotland.com/. Accessed 4 Oct 2025
27. Green Car Congress. Electric truck ready for dynamic wireless charging on public roads in Sweden: Smartroad Gotland. https://www.greencarcongress.com/2020/02/20200226-electreon.html. Accessed 4 Oct 2025
28. Siemens Mobility. eHighway—electrification of road freight transport. https://www.mobility.siemens.com/us/en/portfolio/rail-infrastructure/electrification/ehighway.html. Accessed 4 Oct 2025
29. Civil + Structural Engineer Magazine. First U.S. eHighway demo running in California. https://csengineermag.com/first-u-s-e-highway-demo-running-california/. Accessed 4 Oct 2025
30. Innovations Report. Siemens trials eHighway system for electric trucks in California. https://www.innovations-report.com/agriculture-environment/ecology-the-environment-and-conservation/siemens-tests-ehighway-system-in-california/. Accessed 4 Oct 2025
31. MDPI. An overview of dynamic inductive charging for electric vehicles. https://www.mdpi.com/1996-1073/15/15/5613. Accessed 4 Oct 2025
32. Stellantis Media. "Arena del Futuro" demonstrates capability of dynamic inductive recharging technology for electric vehicles. https://www.media.stellantis.com/em-en/corporate-communications/press/arena-del-futuro-demonstrates-capability-of-dynamic-inductive-recharging-technology-for-electric-vehicles. Accessed 4 Oct 2025
33. Siemens Mobility (2024) eHighway technology overview and demonstrations. https://www.mobility.siemens.com/. Accessed 4 Oct 2025
34. Fraunhofer Institute for Transportation and Infrastructure Systems (IVI) (2024) Hybrid ERS integration for smart cities. https://www.ivi.fraunhofer.de/. Accessed 4 Oct 2025

35. Cision. Installation of the world's first wireless electric road for trucks and buses begins today—Visby, Smartroad Gotland. https://mb.cision.com/Main/18921/2959916/1140125.pdf. Accessed 4 Oct 2025
36. Discovery Journals. Static and dynamic wireless charging of electric vehicles using inductive coupling. https://www.discoveryjournals.org/engineering/current_issue/2023/v20/n53/e25ije1656.pdf. Accessed 4 Oct 2025
37. Indian Institute of Technology Madras (2025) Dynamic charging for EV corridors in India: pilot results and technical framework. https://www.iitm.ac.in/. Accessed 4 Oct 2025
38. ResearchGate. Driving range extension of EV with on-road contactless power transfer—a case study. https://www.researchgate.net/publication/254017791_Driving_Range_Extension_of_EV_With_On-Road_Contactless_Power_Transfer-A_Case_Study. Accessed 4 Oct 2025

Chapter 4
Grid Integration and Energy Management in Electric Road Systems (ERS)

4.1 Introduction

4.1.1 Defining Grid Integration in the ERS Context

Electric Road Systems (ERS) are a breakthrough in the electrification of transport beyond the concept of the immobile charge of vehicles parked to the idea of the mobile charge of vehicles on the move. This conversion requires an extensive and elaborate connection with the electrical grid. In the context of ERS, grid integration refers to the overall system of physical infrastructure, power electronics, communication networks, and control systems that regulate the real-time two-way exchange of energy and data between roadways and the power grid. An ERS is more than a large passively powered electrical load: it is a multi-faceted, geographically dispersed, cyber-physical system made up of 5 major subsystems, (1) the electricity supply, including grid connections and substations; (2) the road infrastructure per se, which contains the power transfer equipment; (3) the power transfer system, which might be catenary lines, in-road rails, or inductive coils; (4) the road operation system, which manages energy flow and the ability of the vehicles to access it; and (5) the vehicles equipped with compatible power receivers [1].

The magnitude of such a load is high. The peak power demand of a single ERS corridor may be tens of megawatts, which may change greatly over time and space depending on the traffic flow. Otherwise, this demand may result in extreme stress on local distribution networks, which causes voltage instability and speedy wear of grid equipment like transformers. Therefore, successful grid integration is not a fringe benefit and is an essential condition of the effectiveness, scale, and trustworthiness of ERS. It transforms the ERS into a potential grid liability into an interactive and dynamic asset that can be involved and contribute to the grid operations. This movement will be based on the transformation of the traditional power grid to the Smart grid, an intelligent and responsive grid [2].

R. Agnihotri, *Electrified Road Systems (ERS) for Emerging Asia*, Springer Tracts on Transportation and Traffic 24, https://doi.org/10.1007/978-3-032-16927-3_4

4.1.2 The Importance of Real-Time Power Exchange

The ability to have bi-directional and real-time power exchange is the essence of modern grid integration. To an ERS, this implies moving past a unidirectional energy transfer between the grid and the vehicle (Grid-to-Vehicle, G2V) to allow the possibility of power moving from a vehicle to the grid (Vehicle-to-Grid, V2G). This two-way communication, enabled by state-of-the-art power electronics and communications protocols, makes the ERS and vehicles connected to it a giant mobile energy storage system.

Real-time power exchange allows an ERS to supply a package of useful ancillary services to the grid operator. Vehicles can temporarily slow down their charge rate or even momentarily put some energy back into the grid during periods of peak demand or high grid stress to assist in stabilising the frequency and voltage. It is made up of the reverse, however: in times of surplus renewable energy production (e.g. high output of the sun at midday), the ERS can draw a greater load, effectively consuming any excess energy that would otherwise be curtailed. This two-way interaction is essential in balancing the intermittency of the variable renewable energy sources and improving the grid stability. The real-time modulation of power consumption and injection has enabled the ERS to enter energy markets, which generate new sources of revenue that can be used to recoup the high cost of the capital infrastructure.

The functionality itself is not just an imagined advantage, but an inevitable development. The power electronics of ERS infrastructure, which is made up of a large quantity of inverters and converters that populate the roadway, are not simply components to deliver power passively. The latest, grid-sensitive power electronics have the ability to do sophisticated tasks such as reactive power compensation and voltage regulation. This implies a paradigm re-conceptualisation of the ERS itself. The network of power converters along an ERS corridor can be designed to operate as a distributed Static Synchronous Compensator (STATCOM) instead of being perceived as an issue in the form of a large load that needs to be managed. The ERS plays an active role in this role to ensure that the grid is supported through the regulation of local voltage profiles as well as the enhancement of power quality. This transforms the ERS into an active or grid-forming asset rather than a passive or grid-following one, which improves the resilience and stability of the distribution network to which it is connected. This fact means that the design of ERS infrastructure should take into consideration highly developed inverter control strategies since the beginning of its design, and the economic feasibility of its implementation should consider possible revenues of the receipt of these important grid services.

4.1.3 *The Role of Smart Grids and Distributed Energy Resources (DER)*

ESR demands a high level of interaction that is real-time, something that can only be achieved under the Smart Grid. A Smart grid is a modernisation of the traditional electrical grid that combines the latest technologies in communication, sensing and control to allow the flow of electricity and information in both ways with utilities and consumers. The major enabling technologies of the Smart Grids that are vital in ERS are:

- **Advanced Metering Infrastructure (AMI)**: Real-time data on power usage is provided by this infrastructure, which is required in billing, load forecasting, and demand response programmes.
- **Demand Response (DR) Systems**: Allow grid operators to signal consumers to reduce or shift their electricity usage during peak periods in exchange for incentives, a crucial tool for managing ERS load [3].
- **Integration of Distributed Energy Resources (DERs)**: Smart grids must support and coordinate a group of diverse DERs (also known as ERS), but also rooftop solar panels, local wind turbines, and battery storage systems [3].

In such an ecosystem, an ERS is a big, manipulable DER. It is possible to control its energy consumption intelligently and make it orchestrate with other DERs to effectively optimise the performance of the local grid. An example is that the charging request of an ERS can be planned to align with the times when the local solar farms are most productive and utilise renewable energy to the maximum, and use fossil-based power plants minimally. The future of energy lies in this coordination, which is a decentralised, resilient and sustainable future.

4.1.4 *Chapter Roadmap*

The chapter provides an in-depth discussion of grid integration and energy management policies that should form the basis of the successful implementation of Electric Road Systems (ERS). The analysis has a logical flow, and it starts with the basic composition of the electrical structure and goes to advanced and smart control systems. Section 4.2 clarifies how power flows and the grid structure of ERS, outlining the most important power electronic elements and the possibility of bidirectional energy exchange. Section 4.3 touches upon the important issue of dynamic load balancing and demand-response mechanisms to control the large and unpredictable ERS load. Section 4.4 explores the synergistic application of ERS and renewable energy sources and analyses the utility of microgrids and energy storage in maintaining a stable and sustainable power supply. In Sect. 4.5, the economic aspect is considered, and dynamic pricing models and tariff schemes that could encourage efficient energy consumption are addressed. Section 4.6 discusses how the concept of artificial intelligence and edge computing can be used to transform how predictive energy management and real-time grid coordination can be made possible. Section 4.7 provides a more realistic evaluation of the technical, operational and security-related issues related to grid integration and presents future reduction measures. Section 4.8 is futuristic, exploring such evolving trends as decentralisation of the grid, energy trading based on blockchain, and Vehicle-to-Everything (V2X) networks. The last, but not the least, Sect. 4.9 ends with a summary of the symbiotic relationship between ERS and the grid and the direction to go in the full integration of an intelligent energy and transportation ecosystem.

4.2 Power Flow and Grid Architecture in ERS

The physical interface of an Electric Road System to the power structure consists of a multi-step power delivery paradigm to modify the high-voltage electricity of the transmission system into the particular form necessary at the power transfer interface at the roadside. Such architecture should be strong, efficient and controllable to handle high power requirements and dynamic requirements of ERS loads.

4.2.1 Grid Topology for ERS

An ERS is represented by a grid topology, which is hierarchical, reducing voltage and transforming the power through multiple stages between the bulk power system and the vehicle.

- **Transmission Grid Connection**: A large-scale ERS deployment can generally demand a specific connection to the high-voltage (HV) transmission system, which can be at levels of 135 kV or more. This guarantees a reliable and large capacity source of power, and thus, the effect on the distribution network is reduced in the region.
- **Dedicated Substations**: Dedicated traction power substations (TPSS) are fitted at the point of connection. These substations contain very large transformers to reduce the HV transmission voltage to a medium voltage (MV) voltage, usually in the range of 10–50 kV AC. There are also switchgears, protection relays and control systems in these substations to control the flow of power and to maintain safe operation. In the case of a catenary-based ERS, which could be directly run at 25 kV AC, substation design is comparable to that of electric railways.
- **Medium Volt Feeders**: Interconnecting the substation is a system of MV feeders, which are parallel to the electrified roadway, and may be buried underground or placed in ducts along the right-of-way. These feeders are part of the support of the ERS power distribution system, which will supply power to the roadside cabinets at consistent distances.

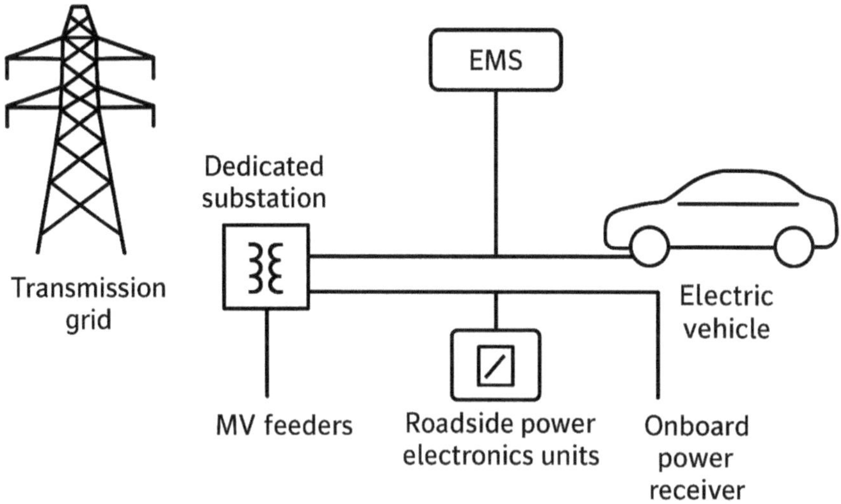

Fig. 4.1 Provides a schematic representation of this multi-stage interconnection, illustrating the flow of energy from the transmission grid to the vehicle

- **Roadside Power Electronics Cabinets**: Roadside power electronics cabinets are cabinet-based power electronics spaced at a minimum of several kilometres to a few hundred metres, again depending on the power demand and ERS technology. These units carry out the last conversion of power. In conductive rail or inductive ERS, rectifiers are usually fitted in these cabinets to convert the MV AC power to a lower DC voltage (e.g. 600–1500 V DC) or inverters that run at high frequencies to produce the AC magnetic field needed to transfer power wirelessly (e.g. 85 kHz). Such units play a vital role in regulating the supply of power to specific sections of the roads in real time (Fig. 4.1).

Certain parameters of this architecture, including voltage levels and power ratings, are dependent on the selected ERS technology. Table 4.1 provides a summary of the nominal specifications of catenary, conductive and inductive systems, giving a broad overview of their unique grid interface requirements. Such consolidation of parameters is necessary to preliminary engineering design, grid impact assessments and economic feasibility studies because it makes clear the infrastructure requirements peculiar to any individual technology. An example is that in catenary systems, it is possible to operate directly on an MV AC supply, which simplifies the electronics required in the roadside but in-road systems require a wider network of local power conversion units, which makes it more expensive, complex and less efficient.

Table 4.1 Power specifications and grid connection parameters for different ERS technologies

ERS technology	Typical operating voltage (roadside)	Typical power transfer per vehicle (kW)	Grid connection voltage (feeder)	Key power electronics
Catenary (overhead)	12–25 kV AC; ~750 V DC	200–600 (heavy duty)	25–135 kV AC	Substations, autotransformers
Conductive (in-road rail)	600–1500 V DC	150–300 (heavy duty); 50–100 (LDV)	10–25 kV AC	Roadside rectifiers, inverters
Inductive (wireless)	High-frequency AC (20–85 kHz)	50–200 (heavy duty); 20–50 (LDV)	10–25 kV AC	Roadside high-frequency inverters

Source Synthesised from [4]

4.2.2 *Power Conversion, Voltage Regulation, and Power Flow Control*

Sophisticated power electronic converters control the interface of the AC grid and the Electrified Roadway System (ERS). These devices are the ones that change electrical power into the necessary form, control the voltage and current, and ensure the high quality of power.

- **AC/DC and DC/AC Conversion**: In conductive rail systems, which have historically used DC, roadside units are using high-capacity rectifiers to convert the received feeders of AC supply into a constant DC voltage. In systems where support of Vehicle-to-Grid (V2G) features is provided, however, these units are also required to include inverters that can convert vehicle-generated DC power into grid-synchronised AC power.
- **Voltage Regulation and Power Factor Correction**: An ERS can bring about voltage sags and swells on the distribution feeders due to the large and variable loads and surges it causes. The control systems of the power-electronic converters, which interface to the grid, actively control the output voltage to ensure it is within the statutory limits (e.g. the $\pm 5\%$ nominal). They are also used to control reactive-power exchange with the grid to maintain a high power factor (generally close to unity) and hence a minimum of electrical losses and a maximum utilisation of grid capacity.
- **Harmonic Mitigation**: Power converters that are made of non-linear loads can cause harmonic currents in the grid. Modern ERS converters use modern switching techniques, including Pulse Width Modulation (PWM) and use active or passive filters to reduce the injection of harmonics and meet strict standards of power-quality requirements, including IEEE 519 [5].
- **Grid Synchronisation**: To allow any operation of power being inserted into the grid (i.e., V2G), the ERS inverters are required to synchronise their output AC voltage with that of a grid in terms of frequency, phase and amplitude. Phase-locked loop (PLL) control systems are used to achieve this synchronisation. The loss of synchronisation may result in huge fault currents and may cause the local grid to become unstable; fault-tolerant and high-strength synchronisation is an essential safety and operational requirement.

4.2.3 *Bidirectional Power Flow (V2G) in ERS*

The Vehicle-to-Grid (V2G) concept converts electric vehicles into active, distributed energy resources, although commonly considered in the framework of stationary charging [6]. V2G may be applied to the dynamic ERS environment, though. An ERS-V2G system allows vehicles passing through the electrified road to both receive power and contribute power to the road infrastructure, which is in turn connected to the broader electrical grid.

This possibility opens up a vast potential for grid support. A large battery pack would have a huge amount of energy-storage capacity, and a fleet of thousands of heavy-duty trucks each would amount to a large amount of energy-storage capacity as a whole. Through the coordination of the charging and discharging of these mobile batteries, an ERS operator will be able to offer useful services to the grid [7].

- **Peak Load Levelling (Peak Shaving)**: During hours of peak system demand, the vehicle fleet can reduce its collective charging load or discharge power to the grid, reducing the strain on conventional power plants and transmission infrastructure [8].
- **Frequency Regulation**: Battery inverters have rapid response times that allow the vehicle fleet to inject or absorb brief bursts of power to stabilise grid frequency, which is are important service to keep the grid reliable.
- **Renewable Energy Integration**: Vehicle fleet may be considered a large distributed battery, capturing the surplus production of solar or wind energy when it is abundant and releasing it when the demand is high or the output of renewable energy is low. This helps to smooth the intermittency of renewable energy and improves the overall utilisation [6].

ERS-V2G is an implementation that requires a number of important technical components. Cars will have onboard chargers that have bidirectional features for handling a two-way power supply. The roadside power electronics should also be bidirectional, i.e. capable of both rectification (G2V) and inversion (V2G) [7]. Lastly, an advanced communication and control system is needed to coordinate and allocate the overall capacity of the vehicle fleet in response to grid operator signals, without impairing the original transportation mission of that particular vehicle.

4.3 Load Balancing and Peak Demand Management

One of the main problems of ERS integration in the power grid is how to manage the huge and highly fluctuating electrical load of the ERS. Uncontrolled, this peak demand is predictable but severe, reaching up to 20% relative to normal traffic, and is likely to cause severe voltage instability, rapid transformer ageing, and even necessitate grid reinforcements at a high cost [9]. Simulation-based studies of distribution feeders suggest that the sudden, synchronised demand increase associated with EV charging can cause a 20% relative to normal increase in peak demand, causing serious strain on the infrastructure. As a result of this, higher-order load-balancing and peak-demand-management mechanisms are not only optimisations but also critical operation demands of a scaling ERS.

4.3.1 *Dynamic Load Balancing (DLB)*

Dynamic Load Balancing (DLB) is a real-time energy management method that assigns the available electrical capacity to different segments of the Electrified Road System (ERS) network in an incentive-friendly fashion. Instead of enforcing an a priori limit on the amount of power per roadside section, a DLB system continuously measures the cumulative capacity of the grid connection as well as the force of passing vehicles and dynamically distributes this power according to need without violating grid capacity limits [10].

An integrated DLB system is based on three major elements:

1. **Real-Time Data Analysis**: The system takes in information provided by various sources, such as smart meters that are at the point of grid connection (to measure the total load), traffic sensors (to estimate the number of vehicles in the traffic), and the communication of vehicles (to demand power). This data stream is processed by sophisticated analytical algorithms, which build a dynamic supply and demand representation throughout the network.
2. **Smart Charging Algorithms**: The DLB system is based on central algorithms that make allocation decisions. These are simple rule-based systems to more advanced optimisation procedures. One of the most commonly used paradigms is the so-called water-filling algorithm, which analogically fills the provided grid capacity into the containers of vehicle demand, thus fulfilling the most urgent or the most high-priority one until the capacity is filled. The algorithms could give priority to charging based on the state-of-charge (SoC), destination or price of electricity in a vehicle.
3. **Communication Infrastructure**: Good DLB assumes a high-bandwidth, low-latency communication system between the central management system, road-side power electronics, and vehicles. Interoperability, which helps the system to direct specific control commands to each charging segment to adjust its power output in real time, is supported by such standardised protocols as the Open Charge Point Protocol (OCPP).

With the implementation of DLB, an ERS operator will increase the use of the available grid connection, operate a larger fleet without having to install expensive infrastructure, and reduce the risk of overloads, which may endanger grid stability.

4.3.2 *Demand Response (DR) and Time-of-Use (TOU) Scheduling*

Whereas the DLB is the load management within the ERS characteristic limits of capacity, Demand Response (DR) programmes are the response of the ERS to external indications of the grid operator to alter the aggregate load. These programmes are a foundation of smart-grid operation, which provides a versatile resource in supply–demand balancing in the entire power system. There are multiple ways through which an ERS can practice DR:

- **Price-Based DR**: In this model, the ERS operator reacts to the dynamic electricity prices. Time-of-Use (TOU) is by far the most common, where the rate is increased during established peak periods and decreased during off-peak periods. This incentive system encourages the ERS operator, and ultimately the freight operators who are using the system, to redistribute the charging operations to off-peak periods, i.e. outside the night, to a cheaper period. Another more advanced system can use Real-time Pricing (RTP), which allows prices to be changed on an hourly basis, based on the wholesale market conditions and allows finer-grained optimisation.
- **Incentive-Based and Emergency DR**: Under such programmes, the ERS operator signs a contract with the grid operator to be called on to reduce a given load-reduction capacity in a given situation. The grid operator can send out a dispatch signal when the grid is stressed (e.g. when it is in the middle of a heatwave, or when a large power plant suddenly goes offline). The operator of ERS then limits its load (e.g. by temporarily limiting the maximum power of the roadway) and is paid compensation in terms of the service of reliability provided. An example of such an arrangement is the Electric Reliability Council of Texas (ERCOT) Emergency Response Service (ERS) that requires the participants to respond within a specified period, 10 or 30 min.

Figure 4.2 illustrates the control flow for these integrated load management strategies, showing how data from the grid and the transportation network informs the decisions of a central Energy Management System (EMS).

This diagram shows the flow of information and control commands in an integrated ERS energy management system. Inputs from the grid operator (e.g., DR signals, real-time prices) and the road network (e.g., traffic data, vehicle requests) are processed by the EMS. The EMS then executes control actions, such as dynamically allocating power across road segments or curtailing overall load, by sending commands to the roadside power converters.

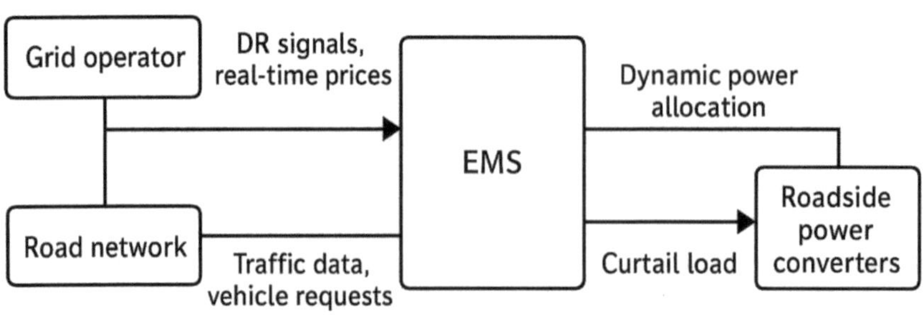

Fig. 4.2 Load balancing and demand response flow diagram

Another major issue that can be linked to simple TOU tariffs is the so-called avalanche effect or rebound peak. This effect occurs when a massive pool of manageable loads, including electric vehicles, is triggered at the same time when the off-peak rate is activated, and the sharp peak of demand is created again, which can be just as harmful as the first one. It has been found through simulations that the effect can be mitigated by randomising the time of charging sessions in the off-peak period, redistributing load and resulting in a peak demand of up to 5% less than with no TOU rates. This highlights the need to have sophisticated and actively controlled charging schemes rather than passive TOU timetables.

These different load-management strategies can be successfully assessed on the basis of a series of key performance indicators (KPIs). Table 4.2 identifies these measures and gives the commonly used target measures, thus providing a standardised framework for evaluating and comparing the performance of different approaches. It helps to make decisions on the system operators and regulators based on data, and promotes the choice of strategies whose value is the highest in terms of cost savings, peak reduction, and grid stability.

Table 4.2 Load balancing and peak management metrics

Metric	Definition	Unit	Typical target value	Relevant strategy
Peak load reduction	The reduction in maximum power demand compared to an unmanaged baseline	MW or %	>15%	DR, TOU, DLB
Load factor improvement	Ratio of average load to peak load over a period. Higher is better	Dimensionless (0–1)	>0.75	TOU, DLB
Peak-to-average ratio (PAR)	The ratio of peak power to the average power level. Lower is better	Dimensionless	<1.5	TOU, DLB
Demand response time	Time taken to curtail load after receiving a grid operator signal	Minutes	<10 min (for ERS-10)	Emergency DR
Cost savings	Reduction in electricity procurement costs due to shifting load from high-price to low-price periods	% or USD/MWh	5–15%	TOU, real-time pricing

Source Synthesised from [9]

4.4 Integration of Renewable Energy Sources

The main reason why transport is being electrified is the reduction of greenhouse-gasses. This goal can be fully achieved when the power that is used to drive the ERS is renewable and is also clean. The inclusion of variable renewable energy sources (VRES), including solar and wind energy, therefore, forms a vital part of a sustainable ERS ecosystem. Nevertheless, the natural intermittency of these sources, solar power only in the daytime, and wind power with stochastic characteristics depending on the weather conditions, pose a significant problem to grid stability, which should be managed by means of intelligent design of the system [11].

4.4.1 Architectures for Renewable Integration and the Role of Microgrids

The incorporation of renewable energy into the ERS power supply is possible using a number of architectural models. Wind and solar farms on a large scale can be linked to the high voltage transmission grid, resulting in the delivery of green energy to the bulk power system, on which the ERS obtains its power [12]. Although this type of arrangement helps with the general decarbonisation, it does not help to address the local grid issues directly or increase the local resilience.

The more developed and robust architecture is the one that implies the implementation of local microgrids. A microgrid is a geographically limited grouping of sources of power generation and electrical loads that can be connected simultaneously with the primary grid and still be able to act independently in an island mode in the event of a failure of the primary grid. In the case of an Electric Road System (ERS), such a set-up may be the placing of photovoltaic (PV) arrays along transport corridors or on the roofing of adjacent logistics hubs, or built into noise reduction structures. The distributed generation units may be coupled with the energy storage facilities to create an autonomous power nexus of a specified part of the ERS [13].

This microgrid approach offers several key advantages:

- **Enhanced Resilience**: During the time of a system-wide microgrid outage, a microgrid-based ERS section can continue to operate, thus ensuring the availability of essential freight routes.
- **Reduced Grid Congestion**: Microgrids make the system less congested, as with small-scale generation, the amount of energy that has to be moved over long distances is reduced, and this reduces the congestion on the main grid.
- **Optimised Renewable Energy Use**: Local generation can be visibly adjusted to a particular demand of ERS, resulting in optimisation of losses and minimisation of clean energy use.

This close linkage of generation and load at the same time contributes to a healthy economic drive. The ERS, with its enormous, predictable and persistent demand for electricity, is an ideal anchor tenant to fledgling renewable energy projects. A long-term Power Purchase Agreement (PPA) with the ERS operator could be contracted with the developers of solar or wind generation facilities, which can ensure an effective revenue stream that is valuable in the acquisition of capital. The result of this symbiotic relationship is the creation of a self-perpetuating cycle whereby the establishment of ERS infrastructure directly drives the construction of new renewable generating capacity, therefore accelerating the decarbonisation of the transport and energy sector.

4.4.2 Energy Storage Solutions for Mitigating Intermittency

The main issue associated with the integration of variable renewable energy sources (VRES) is the way to handle their variability. The key enabling technology to close the temporal gap between non-generative intermittency and changeable demand in ERS is the energy storage systems (ESS) [3]. Various storage technologies have varying temporal scales:

- **Battery Energy Storage Systems (BESS)**: Lithium-ion battery arrays of large scale are ideal for energy time-shifting in timeframes ranging over several hours. A BESS placed in connection with a solar-powered ERS micro-grid can store the excess solar energy in the middle of the day and later release it during the late afternoon and evening traffic conditions, which will match the demand. ESS capacities suggested to peak-shave during ERS use fall between 6 and 45 MWh, depending on the traffic and the target degree of grid independence.
- **Supercapacitors and Flywheels**: These kinds of technologies store relatively lower amounts of energy, but facilitate a comparatively fast charge and discharge cycle. They are especially suitable for the provision of rapid response ancillary services like frequency restraint, and also the mitigation of transient power fluctuation as a consequence of individual vehicles linking to and disconnecting the ERS.

Figure 4.3 presents a schematic of a hybrid grid architecture that combines an ERS segment with a local microgrid incorporating solar PV, wind generation, and battery storage, all while maintaining a connection to the main utility grid for backup and energy trading.

4.4.3 Case Study: Applying Solar Microgrid Principles to ERS

The biggest ERS implemented with a particular microgrid is still in pilot mode, but some experience can be gained based on the existing microgrid deployments

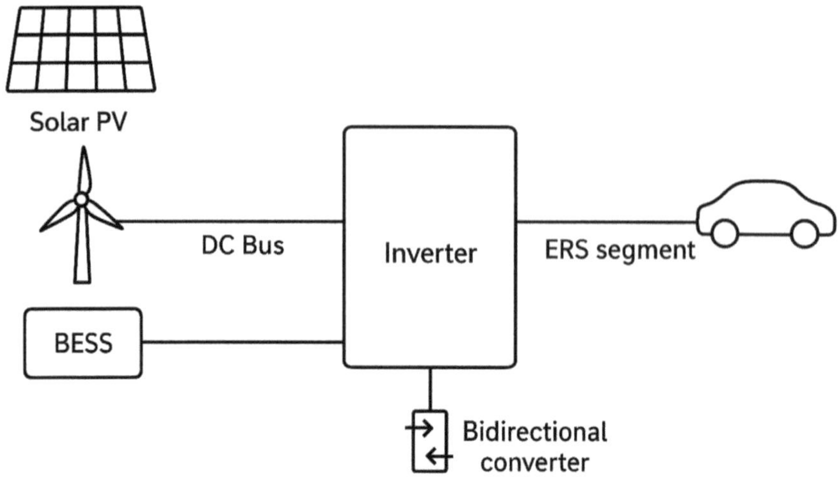

Fig. 4.3 ERS–renewable hybrid grid layout

that accommodate similar high-power and variable loads, e.g., electric vehicle (EV) charge depots. One good example is the solar microgrid installed on the Direct Relief headquarters in Goleta, California, a region with a favourable solar potential profile as well as grid vulnerability.

The Direct Relief centre, which serves as an emergency humanitarian distribution centre, installed a microgrid to ensure that the facility remains operational during regular grid failures. The system includes a 320 kW PV array, a 676 kWh lithium-ion battery system and a backup diesel generator. The load profile of the facility comprises a large number of EV chargers, which makes it an ideal proxy of an ERS-based logistics hub.

The microgrid working scheme provides a plan that can be used during the implementation of ERS:

1. **Solar Generation**: On sunny days, the loads to the facility are provided by the PV array.
2. **Energy Storage**: The battery system is charged by the excessive production of solar energy.
3. **Peak Demand Shaving**: During the afternoon (when the sun is setting and the facility demand is high), but the facility is not producing much power (which coincides with high utility pricing), the battery will be discharged to fill the loads, eliminating the necessity to purchase expensive grid power.
4. **Resilience**: In the event of the main grid failure, the microgrid automatically goes offline, using the PV array and battery to provide power to critical loads indefinitely.

After a year of microgrid operations, 57% of all energy consumed by the facility was attributed to the microgrid. The combination electricity rate was slightly above (about 3% premium) that of absolute utility-only dependency; however, this price was considered a prudent investment considering the priceless strength the utility was given. Scaling of this model to an ERS scenario can be used to imply the feasibility of the implementation of a series of these types of solar-plus-storage microgrids at logistics centres and rest stops along a highway traffic network. Such microgrids would serve dual functions of providing clean energy to dynamic charging opportunities on the major ERS and charging stationary high-speed chargers at the hubs, which would add to the overall energy security of the transportation system. Research into the future energy outlooks in Germany and California has highlighted the importance of the fact that intelligently controlled EV charging, including ERS, must contribute to the absorption of the large-scale generation of wind and solar resources common in the regions [14].

4.5 Energy Pricing and Economic Management

The issue of the economic viability of an Electric Road System is inherently dependent on its energy management approach. Electricity cost is one of the major operating expenses of the facility, and the ability of the facility to acquire and use this energy in an economically viable way is critical towards long-term systemic profitability. This is the reason why dynamic pricing models and advanced economic management tools are not only the tools that can be used to reduce costs but also the tools without which the mechanisms of ERS operational behaviour cannot be aligned with the physical and economic parameters of the broader power grid.

4.5.1 *Concepts in Dynamic Energy Pricing*

The conventional electricity tariffs are normally based on a flat rate, in which the price per kilowatt hour is the same regardless of the time of day. This method provides no motivation to shift the load and does not suit massive and flexible loads, e.g., electric vehicle charging stations (ERS). Dynamic pricing plans are developed to reflect the changing price of producing and delivering electricity and thus will give price signals that will induce consumers to decrease consumption when its cost is high (i.e. when it is on the peak demand) and increase consumption when its cost is low (i.e. when it is in high renewable generation). Prominent models include:

- **Time-of-Use (TOU) Tariffs**: It is the most basic type of dynamic pricing. The day is classified into pre-determined periods-on peak, off-peak, and at times a mid-peak or shoulder period, all with set and determined prices. An example here is that a utility would charge a large price between 3 pm and 8 pm during weekdays and a very low price at night and during weekends. This design is easily understandable to the consumers and can be well used to motivate the shifting of deferrable loads, which can be freight truck charging at depots to off-peak directly.

- **Real-Time Pricing (RTP)**: RTP is a model of electricity pricing where the price changes often, usually hourly or even sub-hourly, as the market wholesale price varies in real time. This gives a very granular price signal, which allows optimisation. An ERS fitted with an automated energy management system can use RTP data to continuously adjust its charge load, minimising consumption when prices are high and maximising it when prices are low, thus saving a significant amount of total energy expense.
- **Critical Peak Pricing (CPP)**: CPP is an intermediate type of pricing that is a combination of a standard TOU tariff and extraordinarily high prices during a limited number of critical peak hours or days of the year. These are severe events that are pre-announced by the utility when the grid is experiencing severe stress. The extremely elevated cost creates a great economic motivation for large customers like an ERS to reduce their consumption significantly during these short yet critical seasons, which can avoid blackouts.

4.5.2 Benefits and Models for ERS Economic Management

They can allow the operator of the ERS to become an active participant in the energy market, instead of a passive price-taker, thanks to the adoption of these flexible pricing mechanisms. The advantages are two: the direct cost of operational ERS is minimised, and the operator can offer an excellent flexibility service to the grid [15].

The economic and contractual association between the ERS operator, grid operator and energy suppliers may be designed in several ways:

- **Large Industrial Customer Model**: This is where the ERS operator might be a very large industrial customer who is purchasing electricity either at the wholesale market or in a bilateral contract with a generation company. Under this model, the operator is entirely exposed to price fluctuation but also has the highest possibility to reap the benefits of intelligent load shifting as a reaction to RTP.
- **Aggregator Model**: In this role, it would coordinate the aggregate flexibility of the ERS and its affiliated vehicles and sell this as a service in auxiliary markets, like demand response or frequency regulation markets. The grid operator would pay the operator to have standby capacity, and the actual load curtailment, and this would provide another revenue stream.
- **Public–Private Partnership (PPP) Model**: Since the infrastructure costs are high, and the ERS is a benefit to the people, a PPP model is often taken into consideration. The first construction may be subsidised by a governmental agency, with the day-to-day energy acquisition and operations being conducted by a private operator. The tariff framework of the vehicles using the ERS would need to be developed in a way that it would offset the cost of operations and help in capital investment in the long run.

Table 4.3 Tariff and pricing model comparison in pilot projects and related sectors

Tariff model	Region/country example	Mechanism	Primary goal	Potential ERS application
Feed-in tariff (FiT)	Germany (early renewables)	Guaranteed premium price for renewable energy fed into the grid	Accelerate renewable deployment	Incentivizing ERS-V2G or co-located solar to export power
Time-of-use (TOU)	USA (California), Europe	Static, pre-defined price tiers for peak/off-peak hours	Shift load away from predictable peak periods	Encouraging overnight charging for freight trucks at depots
Real-time pricing (RTP)	Europe (Nord Pool)	Hourly prices based on wholesale market supply and demand	Align consumption with real-time grid conditions and VRES output	AI-driven ERS energy management systems can continuously optimize charging to minimize cost
Auction/tender System	Europe (recent renewables)	Competitive bidding process to set the price for new capacity or energy supply	Drive down costs through market competition	Large ERS operators could participate in auctions to procure long-term energy supply

Source Synthesised from [16]

A comparative analysis of various models of tariffs that have been implemented or suggested in pilot projects and allied energy sectors in Europe and Asia is provided in Table 4.3. This analogy explains various economic processes as being optimised to meet a particular policy goal, which could include increasing the rate at which renewable technologies are implemented or lowering the cost through competition. In the case of ERS, a hybrid solution is probably most useful, whereby entering into a long-term power purchase agreement (PPA), to secure a minimum supply of renewable energy, and at the same time participating in the real-time markets to optimise the costs and offering ancillary grid services.

4.6 AI-Based Energy Routing and Edge-Grid Coordination

The customary method of energy management that relies on a set of rules and reactive controls fails to streamline a system as dynamic and intricate as an ERS. The combination of artificial intelligence (AI) and machine learning (ML) is a landmark innovation that allows shifting the control system towards a predictive and adaptive one. This development enables the ERS to react to any prevailing condition of the grid but also predict the future and optimise its operations in advance [17].

4.6.1 The Role of AI/ML in Predictive Load Management

Proper prediction forms the basis of smart energy management. AI and ML procedures are excellent in detecting sophisticated patterns and non-linear connections in massive data sets, and hence are best applied when estimating the energy requirement of an ERS.

- **Load Forecasting**: Long Short-term memory (LSTM) and Gated Recurrent Unit (GRU) networks are deep-learning models that have proven to be successfully applied to time-series forecasting. These models are capable of predicting the ERS load profile in the upcoming hours and days with very high accuracy by training on historical data that has an input of the traffic flow, time of the day, weather conditions, and special events. This is also possible to allow the system operator to predict peak demand periods, plan energy purchase, storage dispatch and demand-response activities ahead of time [18].
- **Renewable Generation Forecasting**: The AI models can also be used to predict the performance of co-located solar and wind resources, given weather prediction and performance history. This helps the ERS to optimise its charging load and renewable energy.

4.6.2 Advanced Algorithms for Energy Routing and Control

With precise forecast inputs, advanced artificial-intelligence-powered algorithms can determine the most optimal approach to utilise real-time power dispatch, often called energy routing, using a network, and both coordinate vehicle charging and vehicle-to-grid (V2G) activities.

- **Reinforcement Learning (RL)**: Reinforcement Learning (RL) is a potent concept whereby an AI agent will train to optimise the decision-making process through trial and error interactions within a dynamic environment. When used in the context of an electric road system, an RL agent has the ability to learn to control the overall energy subsystem, including grid draw, battery dispatch, and the distribution of power to roadway segments. The reward function of the agent can be designed to achieve the following outcomes: minimising electricity spending, maximising the use of renewable energy sources, and supplying grid-stability services. The RL frameworks with a specific focus on Deep Q-Networks (DQN) have proven themselves to be able to acquire advanced control policies that are far more successful than traditional approaches. Empirical analysis results demonstrate that RL-based routing and energy management can save upwards of 20% of the consumption compared to classical approaches, by adapting dynamically to real-time grid conditions and traffic problems [19].

- **Predictive Optimisation**: Predictive optimisation is a combination of machine-learning predictions and mathematical optimisation programs. The system uses forecasted load and generation profiles to minimise an optimisation problem that would produce the most cost-efficient vehicle charging and energy storage release schedule over a future horizon (e.g. the next 24–39 h).

These high-end AI methods lead to a primal overlap of two previously distinct fields: logistics in the transport sector and electrical-grid activities. A clever ERS management system is unable to optimise every sector separately. As an example, the path through a grid area with extreme congestion and high electrical costs may be the shortest, but an extra mile more would put the truck on a path through an area with an abundant supply of inexpensive solar power. As a result, the concurrent co-optimisation of vehicle routing and energy dispatch by the AI control architecture is necessary. The resulting optimum vehicle path is a multifaceted algorithm of the travel duration, traffic dynamics, dynamic energy costs, and local grid requirements, which is an indicator of a paradigm shift towards a truly integrated intelligent transportation and energy system.

4.6.3 Edge Computing and the Internet of Energy (IoE)

Several ERS control measures, e.g. frequency-response control or in-vehicle power negotiations, need to be implemented in milliseconds. Dependence on a centralised, cloud-based AI framework for making these decisions creates prohibitive latency. Addressing this challenge in latency, edge computing places the resources and AI models close to the data sources, that is, in the cabinets of roadside power-electronics or in the roadside communication units. These edge nodes can do real-time processing on data and provide local control instructions with minimal latency. As an example, an edge device may dynamically control load balancing of one highway segment or coordinate V2G replies in a group of trucks (a platoon). The main cloud service is dedicated to tasks that are not as time-sensitive, such as long-term forecast, fleet-wide optimisation and training AI models. This distributed intelligent architecture is a fundamental component of the new Internet of Energy (IoE), a framework that envisions all the energy-consuming and energy-producing devices, vehicles, chargers, solar inverters, and smart appliances to be connected and coordinated in an intelligent network. This multilayered AI coordination architecture is shown in Fig. 4.4.

Vehicle-to-Everything (V2X) communication also makes the ecosystem possible. V2X is a continuation of V2G, in which direct communication is provided between vehicles and various other entities, such as traffic lights (V2I), pedestrians (V2P), and other vehicles (V2V). It is through this enhanced exchange of data that a more

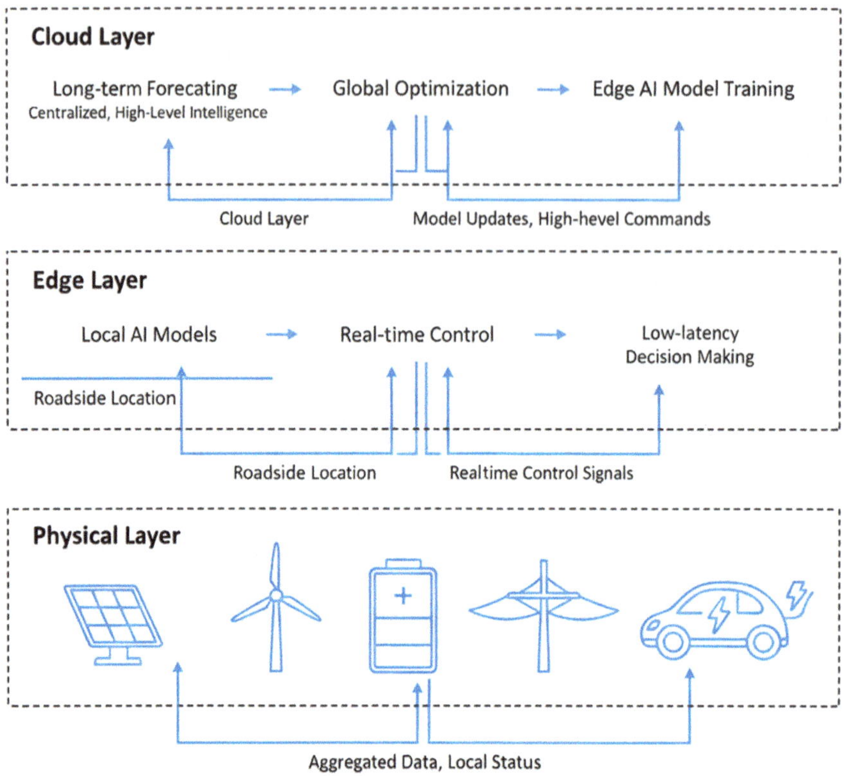

Fig. 4.4 AI AI-based grid coordination architecture

advanced level of coordination is achieved, whereby energy-management decisions are coordinately intertwined with the traffic-flow optimisation as well as safety applications.

4.7 Challenges and Reliability Issues

Although there is enormous potential in the integration of ERS with a smart, renewable-powered grid, the implementation process is marred with huge technical, operational, economic and security challenges. These challenges have to be proactively mitigated to ensure the reliability, safety, and long-term viability of the electric road systems.

4.7.1 Technical Challenges

- **Grid Synchronisation and Stability**: An ERS has thousands of power-electronic inverters spread out along a corridor, which all have to work in perfect synchronisation with the main electrical grid. Synchronisation, especially when the grid is perturbed, e.g. by voltage sags or frequency variation, is a complicated control issue. A loss of synchrony may cause high fault currents, equipment damage and disastrous cascading failures that can destabilise the local grid.
- **Power Quality Degradation**: The extensive installation of power-electronic converters is a serious risk to the quality of power since it is a nonlinear load. These converters have the potential to add harmonic currents into the grid and alter the voltage waveform. Harmonic distortion may result in overheating, failure of the transformer and motors, nuisance tripping of the protection devices and interference with sensitive electronic devices. Even though the harmful impact of harmonic injection is minimised in modern inverters, the cumulative impact of thousands of harmonic injection sources requires careful system-level analysis and may require the fitting of active harmonic filters [20].
- **Power Losses**: Because of the multi-level conversion of power in an ERS, from high-voltage alternating current to medium-voltage alternating current, then to direct current or high-frequency alternating current at the roadside, the conversion process has electrical losses at every step. Such accumulated losses reduce the overall wall-to-wheel efficiency of the system and should be meticulously controlled by the choice of high-efficiency elements of the system and optimised system design.

4.7.2 Operational and Economic Challenges

- **Reliability and Resilience to Blackouts**: The working capability of an ERS depends on the availability of a constant power supply. A single rupture to the main power supply might leave large stretches of an electrified road system out of commission, thus paralysing traffic and causing massive traffic congestion. It follows that a strong level of resilience is essential, which can be achieved by introducing redundant connections of the feeders, local energy storage units, and the ability of single segments to work independently as microgrids during a system outage.
- **Maintenance and Repair**: The physical facilities that constitute an ERS, which include overhead cabling, in-road rails and roadside electronic elements, are vulnerable to the harsh environment and wear and tear caused by repetitive vehicular traffic. Making maintenance and repair work on such equipment, especially the parts that are buried on the road, presents significant logistical issues and safety risks, which often result in a lane closure and subsequent traffic jams.
- **High Capital Investment**: The first capital expenditure incurred in the course of building an ERS, as well as necessary grid upgrades, is also one of the strongest

obstacles to implementation. This is in addition to the road infrastructure (estimated €1.7–3.1 million per km), new distribution feeders, dedicated substations, and power-electronic apparatus. It is a great economic challenge to finance such high-scale infrastructure projects.

4.7.3 Cyber-Physical Vulnerabilities

An Electric Road System (ERS) is a classical cyber-physical system since it is a massive system, interconnected physically with digital control and communication networks. Such interconnection makes it one of the major attackable areas, with potential outcomes of cyberattacks being disastrous. The scope of the attack is extensive and includes not only the onboard systems of the vehicle but also roadside communication units, the central energy management system, and the control centre of the operator of the grid.

Potential cyber threats include:

- **Denial-of-Service (DoS) Attacks**: A hacker might flood the communication network with malicious messages, thus blocking control triggers to roadside units and the charging operation throughout the entire system.
- **Data Manipulation**: An attacker might interrupt and modify information, including electricity pricing data to commit fraud, or modify sensor data to cause the computer to implement dangerous control actions.
- **Direct Control System Intrusion**: An advanced attacker that manages to get access to the central Energy Management System (EMS) or Supervisory Control and Data Acquisition (SCADA) system of the grid may interact directly with power flow, cut off substations, or cause electrical instabilities, which can physically damage the equipment in the grid, making it a public, safety-threat.

Such threats need to be mitigated with a defence-in-depth cybersecurity strategy that includes network segmentations, encrypted communications, strong access control, continuous monitoring of intrusion, regular security audits and vulnerability analysis. In the case of power systems, the availability of the service is of utmost importance; therefore, the cybersecurity strategies should be tailored in such a way that they guarantee the uninterrupted supply of power.

Table 4.4 provides a systematic overview of these major challenges and describes the relevant mitigation efforts that can be utilised to handle the risks posed by ERS grid integration.

Table 4.4 Grid integration challenges and mitigation measures

Challenge category	Specific challenge	Potential impact	Mitigation measures
Technical	Power quality (harmonics)	Transformer overheating, equipment malfunction, increased losses	Active harmonic filters (AHFs), advanced inverter control (IEEE 1547 std.), grid-forming inverters
Technical	Voltage instability/sag	Disruption of charging, tripping of protective relays	Distributed STATCOMs (using ERS inverters), dynamic voltage support, energy storage systems
Operational	Peak demand overload	Transformer stress, feeder congestion, potential for local blackouts	Dynamic load balancing (DLB), demand response (DR) programs, integrated energy storage
Economic	High upfront investment	Project unviability, slow deployment rate	Public–private partnerships, ancillary service market participation, green bonds, government subsidies
Cyber-physical	Cyberattack on control system	Widespread service disruption, energy theft, physical damage	Defense-in-depth security architecture, encrypted communications, intrusion detection systems, regular vulnerability assessments

Source Synthesized from [21]

4.8 Future Trends and Research Directions

The development of Electric Road Systems cannot be discussed outside of larger tendencies in the field of energy, computing, and communications. The future of ERS grid integration is toward a more autonomous, decentralised and interconnected ecosystem, where the separation between the transportation network, the power grid, and the digital world is more vaguely defined.

4.8.1 Grid Decentralisation and Roadway Microgrids

The historic, centralised power-grid design with remote and large-scale power stations providing energy in one direction to passive customers is being replaced by a more decentralised design. The next generation grid will include systems of connected microgrids that can manage the local generation, storage and loads. ERS has a perfect opportunity to be the support of this change along the transportation lines. A roadway microgrid can be a highway fitted with an ERS, co-located solar generation and battery storage. Such microgrids will be able to improve the resilience of energy in a region by ensuring that transportation and the communities in the immediate surroundings have a consistent supply of power, even when there is a general outage of the bulk-power system.

4.8.2 Blockchain-Based Peer-to-Peer Energy Exchange

With the increasing decentralisation in energy generation, new energy trading strategies are now coming up. The blockchain technology has provided the basis for building secure, transparent, and decentralised peer-to-peer (P2P) energy markets. A future ERS ecosystem may provide a situation in which a vehicle with a fully charged battery and no urgent travel requirements may sell energy to a nearby car that requires a charge. Equally, a segment of an ERS that has surplus generation of solar power can sell the power to another segment that has a high demand. They would be automatically implemented through the method of smart contracts and would eliminate the need to use a central utility to clear and settle such transactions and promote a more active and efficient energy market.

4.8.3 Fully Integrated Vehicle-to-Everything (V2X) Networks

Although Vehicle-to-Grid (V2G) has become a major advancement, further developments are on totally integrated Vehicle-to-Everywhere (V2X) networks, which are made possible by the forthcoming wireless communication standards. A V2X world will have vehicles not only communicating with the grid, but also in a constant dialogue with the whole environment:

- **Vehicle-to-Home (V2H)/Vehicle-to-Building (V2B)**: A truck that is parked at a distribution centre overnight and is charged by ERS would enable a building to operate, thus saving the building electricity expenses.
- **Vehicle-to-Infrastructure (V2I)**: The Car will communicate with road lights and toll gates, and other infrastructures to coordinate traffic and energy use to maximise road use.
- **Vehicle-to-Vehicle (V2V)**: Platoons of linked and autonomous trucks would be able to exchange energy with each other, where a vehicle that leads the pack draws power via the ERS and sends it to the wired vehicles that are behind the first.

4.8.4 Autonomous Grid Balancing with AI and Digital Twins

Human capabilities will not be enough to handle a decentralised energy-mobility system that is V2X-enabled. The grid control of the future will be autonomous and motivated by high-level artificial intelligence (AI).

The key facilitating technology is the so-called digital twin—a high-fidelity, real-time, virtual representation of a physical setup or system. A digital version of the ERS (a digital twin) and the grid around it would be constantly informed with sensor data taken in the real world. This safe virtual world will allow AI agents to be trained in order to create and test very complex control approaches to grid balancing, congestion control and response to faults. By experimenting with millions of possible solutions, these AI agents will be able to learn the best policies that would otherwise take a long time to figure out using traditional engineering techniques. When tested in the digital twin, such autonomous control algorithms can be transferred to the physical system with a very high level of confidence, and such a grid can self-stabilise and self-optimise in real time.

The communication technology will boost the development of these future systems. A 6G network with near-instant communication (less than a millisecond latency) and huge bandwidth will offer the nervous system needed to communicate and coordinate the millions of distributed assets of a smart ERS energy ecosystem. The conceptual roadmap that is illustrated in Fig. 4.5 provides the roadmap of this evolution.

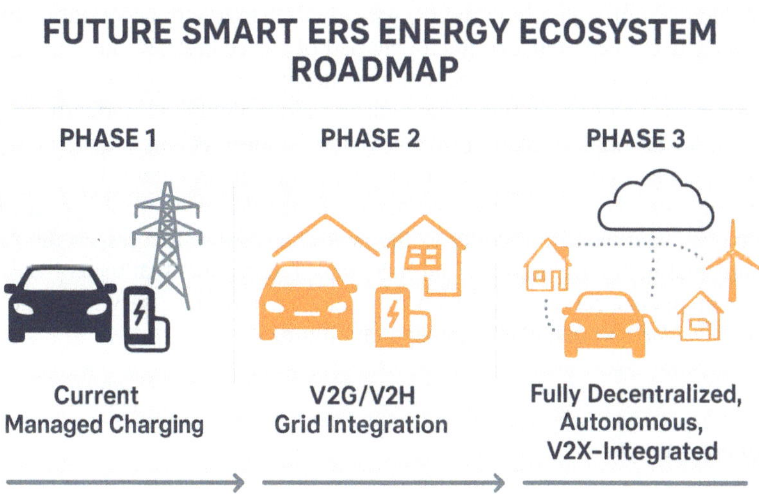

Fig. 4.5 Future smart ERS energy ecosystem roadmap

4.9 Conclusion

Installation of Electric Road Systems (ERS) on a massive basis is not only an issue of transportation engineering, but essentially, it is about the issue of integration of advanced energy systems. As has been shown in this chapter, the association between ERS and the electrical grid is highly symbiotic. Any ERS needs a strong, stable and smart power grid to work with, and an intelligently integrated ERS can be a huge resource, improving the stability, resilience and sustainability of the same grid.

The high-power flow in the high-voltage transmission network to the vehicle has been mapped out with information about the most important power-electronic interfaces and control systems that govern this flow. It has been analysed that the ERS is a daunting load-management problem, but a combination of superior strategies, including dynamic load balancing and demand response, to real-time pricing can turn the problem into grid support. The combination of ERS with renewable energy sources, with the help of local microgrids and energy storage, is the key to the realisation of the final goal of decarbonised transportation. This combination is not only able to provide vehicles with clean power, but also will act as an economic stimulus to increase renewable generation capacity.

The future of ERS is united with the further digitalisation of the energy industry. Use of artificial intelligence, machine learning and edge computing will move ERS management towards a predictive rather than a reactive and autonomous paradigm. Such a shift will drive the integration of energy and mobility management into a common co-optimised field. This transition will be further enhanced by developing emerging technologies, including blockchain-based decentralised energy trading and the creation of full-scale digital twins, which will lead to a transportation network that is an active, integrated, and intelligent part of a decentralised smart grid.

In the end, the technical and economic concepts discussed in this chapter offer a plausible roadmap to a future where energy and mobility are perfectly coupled. However, actual implementation of this vision is not a technical undertaking but one that entails a conducive and progressive policy environment that can inform investment, set standards, and create market forms that can encourage flexibility and innovation. When proceeding to the next chapter that focuses on the critical governmental policy frameworks and roadmaps, one realises that the collaboration between engineers, economists, and policymakers will be necessary to set the stage for an electrified and sustainable transportation future.

References

1. F3 Centre. Electric road systems (ERS). https://f3centre.se/en/fact-sheets/electric-road-systems-ers/. Accessed 7 Oct 2025
2. S. Singh et al. IoT-based energy-efficient architecture for integrated smart grid. https://www.researchgate.net/publication/348107944_IoT_based_energy_efficient_architecture_for_integrated_Smart_Grid. Accessed 7 Oct 2025

3. IEEE-USA. Building an intelligent grid for the 21st century. https://ieeeusa.org/assets/public-policy/white-paper/IEEEUSAWP-BuildinganIntelligentGrid2020.pdf. Accessed 7 Oct 2025
4. Mahmoud A et al. Substation placement for electric road systems. Energies. https://www.mdpi.com/1996-1073/16/10/4217. Accessed 7 Oct 2025
5. Molina MG. A nine-level grid-connected converter topology for single-phase transformerless PV systems. https://www.researchgate.net/publication/260521455_A_Nine-Level_Grid-Connected_Converter_Topology_for_Single-Phase_Transformerless_PV_Systems. Accessed 7 Oct 2025
6. Driivz. What is vehicle-to-grid (V2G)? https://driivz.com/glossary/vehicle-to-grid-v2g/. Accessed 7 Oct 2025
7. Kempton J, Letendre S. Schematic of the vehicle-to-grid (V2G) concept. Reprinted from [10]. https://www.researchgate.net/figure/Schematic-of-the-vehicle-to-grid-V2G-concept-Reprinted-from-10-Copyright-2005-with_fig1_321237450. Accessed 7 Oct 2025
8. Virta Global. Vehicle-to-grid (V2G): everything you need to know. https://www.virta.global/vehicle-to-grid-v2g. Accessed 7 Oct 2025
9. MOST Policy Initiative. EVs electricity demand. https://mostpolicyinitiative.org/science-note/evs-electricity-demand/. Accessed 7 Oct 2025
10. Masoum AS et al. Dynamic intelligent load balancing in power distribution networks. https://www.researchgate.net/publication/275519077_Dynamic_Intelligent_Load_Balancing_in_Power_Distribution_Networks. Accessed 7 Oct 2025
11. IEA. Integrating solar and wind—analysis. https://www.iea.org/reports/integrating-solar-and-wind. Accessed 7 Oct 2025
12. Wang C et al. Microgrid system for electric vehicle charging stations. Front Energy Res. https://doi.org/10.3389/fenrg.2024.1492243/full. Accessed 7 Oct 2025
13. Deilami S et al. Integration of intermittent renewable power supply using grid-connected vehicles—a 2030 case study for California and Germany. Appl Energy 104:666–682. https://ideas.repec.org/a/eee/appene/v104y2013icp666-682.html. Accessed 7 Oct 2025
14. IEA. Executive summary—integrating solar and wind. https://www.iea.org/reports/integrating-solar-and-wind/executive-summary. Accessed 7 Oct 2025
15. Thorson S. The economics of electric roads. https://www.researchgate.net/publication/349411632_The_economics_of_electric_roads. Accessed 7 Oct 2025
16. Bruegel. The economic impact of Trump's tariffs on Europe: an initial assessment. https://www.bruegel.org/analysis/economic-impact-trumps-tariffs-europe-initial-assessment. Accessed 7 Oct 2025
17. Sharma M et al. Power plays: unleashing machine learning magic in smart grids. arXiv preprint arXiv:2410.15423. https://arxiv.org/abs/2410.15423. Accessed 7 Oct 2025
18. Nazari T et al. Deep reinforcement learning for the electric vehicle routing problem with time windows. arXiv. https://arxiv.org/pdf/2010.02068. Accessed 7 Oct 2025
19. Yan S et al. Cyber security in the smart grid: survey and challenges. https://www.researchgate.net/publication/257582416_Cyber_security_in_the_Smart_Grid_Survey_and_challenges. Accessed 7 Oct 2025
20. Georgilakis P et al. A robust synchronization to enhance the power quality of renewable energy systems. https://www.kios.ucy.ac.cy/pv2grid/wp-content/uploads/2020/04/PV2GRID_Journal_1.pdf. Accessed 7 Oct 2025
21. Schneider Electric. Understanding the impact of power quality on electrical systems. https://blog.se.com/services/2025/06/06/understanding-the-impact-of-power-quality-on-electrical-systems/. Accessed 7 Oct 2025

Chapter 5
Government Policy Frameworks and Roadmaps for Electric Road Systems (ERS)

5.1 Introduction

Decarbonisation of the transport sector is one of the most daunting tasks in the process of achieving climate neutrality in the world. Although battery electric vehicles (EVs) have become a major solution to passenger transport, electrification of heavy-duty freight has been a major challenge because of the limitations of battery size, weight, cost, and charging duration. A technologically viable way to overcome these obstacles, at least in the case of long-haul trucking, is provided by Electric Road Systems (ERS)—infrastructure that allows vehicles to dynamically receive power as they travel. Pilot projects throughout the world have shown that several different concepts of ERS can be technically feasible, and may include overhead catenary lines, in-road inductive and conductive systems [1].

However, the shift between successful trial and mass deployment, which is both commercially feasible and technologically viable, is not technologically dominant; it is a policy, governance and strategic investment challenge. The world ERS market is expected to grow considerably, slightly going beyond USD 1.23 billion in 2024 to USD 8.33 billion in 2033, a growth that will only occur due to confident and consistent government intervention. There is a major disparity in regulatory readiness, with the sophistication of ERS technology far ahead of the establishment of the legal, financial and institutional frameworks needed to make it a large-scale project. The stakeholder consultations with the industry leaders have repeatedly shown an agreement that, despite ERS being a central technology in decarbonising road transport, government funding is vital to ERS development to deal with the high investments and risks that a nascent market entails [2].

This chapter conceptualises a Government Roadmap as more than a statement of intent and as a series of multi-year, strategic plans that integrate, harmonise legislative frameworks, financial mechanisms, regulatory standards and institutional coordination. The crux of the matter is that such a roadmap will have to face the basic policy dilemma that governments face between being a catalyst or merely a follower, i.e.

R. Agnihotri, *Electrified Road Systems (ERS) for Emerging Asia*, Springer Tracts on Transportation and Traffic 24, https://doi.org/10.1007/978-3-032-16927-3_5

allowing the market to mature and then implementing the regulations. The experiences of leading countries indicate that in the case of the infrastructure, which requires high capital, like ERS, the catalyst role is necessary to ensure the deployment on a scale and pace matched with climate ambitions [3].

In this chapter, one will be able to get a broad understanding of the policy mechanisms and governance structures that contribute to the development of the ERS in the global context. It starts with a review of the larger policy environment regarding e-mobility and its intersection with the adoption of ERS. Thereafter, it explores the imperative nature of Public–Private Partnership (PPP) models, the necessary regulatory and standardisation framework, and a case study analysis of national roadmaps of Germany, Sweden, the United States and South Korea. Further parts dwell on policy tools and economic inducements, the necessity of institutional coordination and a customised policy structure of India and other Asian economies. The final section of the chapter is the future vision of global harmonisation and how ERS can be aligned with the global sustainability objectives as a means of catalysing this transformative infrastructure by the government.

5.2 Policy Landscape for ERS and E-Mobility

The Electric Road Systems is not developed in a policy vacuum, but rather it is strongly embedded in a wider global trend towards e-mobility and decarbonisation of energy systems. A multi-layered environment of international treaties, domestic goals, and current policies on electric cars and power grid structures determines the policy environment of ERS. The top-level global agreements, the most prominent one being the Paris Agreement, provide the overarching momentum of transport decarbonisation since signing countries are required to take policies that can result in steep emissions cuts, with national policies being informed by the reports and perspectives of international organizations, including the International Energy Agency (IEA) and the United Nations Economic Commission for Europe (UNECE) [4].

On the national level, the development of ERS overlaps with the already established EV policies. Policies like stringent fuel economy and tailpipe CO_2 emissions regulations, which have now expanded to more than 85% of all cars sold globally, have contributed to the growth of the EV passenger vehicle market, acting as an artificial push in that direction [5]. Likewise, Zero-Emission Vehicle (ZEV) requirements, introduced in California and also implemented in China, require manufacturers to produce and market a quota of EVs, creating an artificial impetus to the EV market. Nevertheless, they cannot always be enough to catalyse the ERS implementation directly.

A fundamental mismatch in the design of policies emerges as a critical challenge, as most of the existing e-mobility policies are a consumer-focused response to the consumer-to-business (B2C) market of passenger vehicles. This is in contrast to ERS, especially freight-oriented which is a large-scale capital project (capital-intensive) that functions in a business-to-business (B2B) environment. Its main clients are commercial fleet operators, and its funding mechanism is more like a highway or a power grid than that of a consumer product [2]. A purchase subsidy aimed at making a passenger EV more affordable to an individual is a meaningless effect in financing a multi-million dollar ERS corridor. This requires a policy shift, no longer on consumer-demand incentives but on supply-side policies to benefit infrastructure funding over the long term, grid planning, and development of sustainable B2B business models.

It takes a concerted effort on the part of a variety of institutional stakeholders to develop such a strategy. ERS is at the intersection of multiple ministerial portfolios, which require unprecedented inter-agency coordination [6]. Key actors comprise:

- **Ministries of Transport**: This is the body in charge of the road infrastructure planning, building, maintenance, accessibility of right of way, and the regulation of vehicles.
- **Ministries of Energy**: In charge of the electricity grid capacity, the initiatives of new high-power work, and the design of tariffs, as well as the availability of renewable energy integration.
- **Ministries of Environment and Climate**: Determine the decarbonisation national targets, which establish the strategic requirement of technologies like ERS and frequently supervise the impact assessment of the environment.
- **Ministries of Finance and Economy**: Deal with government financing, formulate economic stimulus, organise the framework of partnership with business, as well as industrial policy.
- **Urban and Regional Planning Authorities**: They are important for zoning, land use, and incorporating ERS corridors into local and regional development plans.

This stakeholder map is complex, and it underscores the need to have specific governance structures that would instil some policy coherence. Unless an integrated strategy is adopted, ERS projects are likely to be derailed by inconsistent regulations, planning discrepancies and institutional inertia. Table 5.1 provides a typology of e-mobility policies and assesses the extent to which they address the specific needs of ERS.

5.3 Public–Private Partnership (PPP) Models

The large capital outlay, extended operation, binding undertakings, and the technological risks involved with Electric Road Systems make their development by either pure public or private funding extremely difficult. This has resulted in the birth of Public–Private Partnership (PPP) as the inevitable formula in funding, building, and maintaining ERS infrastructure. The fundamental concept is that a PPP involves a long-term agreement between a government or agency and a commercial entity to provide a governmental asset or service in which the commercial partner will assume a lot of risk as well as management, which will in turn ensure a better value is delivered to money than conventional public procurement [7].

Table 5.1 Typology of E-mobility policy instruments and relevance to ERS

Policy category	Policy instrument	Primary target	Direct/indirect relevance to ERS
Demand-side incentives	Vehicle purchase subsidies/tax credits	Passenger EVs	**Indirect**: can be adapted for ERS-compatible trucks but is insufficient for infrastructure
	Vehicle registration/ circulation tax benefits	Passenger EVs	**Indirect**: can lower total cost of ownership (TCO) for ERS fleets
Supply-side mandates	CO_2 emission standards	Vehicle manufacturers	**Indirect**: pushes OEMs to develop electric drivetrains, which are a prerequisite for ERS vehicles
	Zero-emission vehicle (ZEV) mandates	Vehicle manufacturers	**Indirect**: accelerates the availability of electric truck platforms that could be adapted for ERS
Infrastructure support	Grants for public charging stations (static)	CPOs, municipalities	**Indirect**: establishes precedent for public funding of charging infrastructure but does not address dynamic charging
	Direct public investment in ERS corridors	Road authorities, tech providers	**Direct**: essential for funding pilot projects and de-risking initial large-scale deployment
	Building codes for "EV-Ready" buildings	Property developers	**Not relevant**: focused on static charging at destinations
Grid integration	Smart charging tariffs (time-of-use rates)	EV owners, utilities	**Direct**: crucial for managing the high, continuous load of ERS to prevent grid instability and minimize electricity costs
	Vehicle-to-grid (V2G) incentives	EV owners, aggregators	**Indirect**: ERS is primarily a grid-to-vehicle (G2V) technology, but grid integration principles are relevant

(continued)

Table 5.1 (continued)

Policy category	Policy instrument	Primary target	Direct/indirect relevance to ERS
Operational incentives	Low/zero-emission zone access	Vehicle operators	**Direct**: can be used to mandate or incentivize ERS-equipped trucks on specific urban or port routes
	Road toll exemptions/ discounts	Vehicle operators	**Direct**: powerful tool to incentivize the use of ERS corridors, thereby ensuring high infrastructure utilization

Various PPP models have varying distributions of duties and risks. Common structures include 11:

- **Build–Operate–Transfer (BOT)**: A third party will finance, construct and run the asset over a concession period with the tendering authority repaying its investment via user charges or availability payments. The ownership is given to the public sector at the expiry of the period.
- **Design–Build–Finance–Operate (DBFO)**: In this model, all the phases of the project are collected together under one private consortium, which designs, constructs, finances and maintains the infrastructure on a contractual basis.
- **Build–Own–Operate–Transfer (BOOT)**: Similar to BOT, but the private entity legally owns the asset during the concession period before transferring it.
- **Joint Venture (JV)**: A government agency and one to three private companies establish a new legal entity to develop the project by sharing the risk, profits, and control.

Nevertheless, when these models are applied to ERS, there is a nuance of a critical note. As versus PPPs relating to mature infrastructure, like toll roads or bridges, where the technology used is standard and the key risk factors are the construction and the demand, ERS projects are essentially all about de-risking and commercialising a new technology. The most significant input of the private partner is not just the timely project delivery, but also exclusive technology and innovation. The strategic goal of the public sector is not merely to acquire an asset but to produce knowledge and prove a new decarbonisation pathway, as well as establish a future market [8]. This will change the risk profile of construction and finance to technology performance, interoperability and future obsolescence. Thus, initial ERS partnerships have been characterised as more of an "Innovation Partnerships": where the public sector acts as a venture capitalist and an initial first customer, the technology risk is jointly borne to trigger a market which would otherwise not appear.

Such an innovation-driven approach can be reflected in the ground-breaking ERS projects on the global level:

- **Germany (Siemens–Government Partnership)**: The German eHighway pilots, including the ELISA project in Hesse, are a case of a government-funded, privately implemented model. The state road authority (Hessen Mobil) manages the project and is funded by the federal ministries (e.g., the Federal Ministry for the Environment) [9]. Siemens, the technology partner, is contracted to plan, construct and maintain the catenary system on an optional basis.
- **Sweden (Electreon–Transport Administration Collaboration)**: The Smart road Gotland project was funded by the Swedish Transport Administration (Trafikverket) via a competitive tender, won by a consortium led by Electreon [10]. The strategic objective of the project was to create knowledge to make future large-scale purchases (e.g., an e-truck of 40 tonnes and an e-bus) [10]. The technology was deployed by Electreon and its partners, with the project being managed by them and the technology being tested.
- **United States (Caltrans/SCAQMD–Siemens Pilot)**: eHighway demonstration project in Carson, California, on the approaches to the Ports of Los Angeles and Long Beach, had a model of a multi-agency public partnership. The US 13.5 million funding of the project was a mixture of different sources of public and quasi-public funding, including the South Coast Air Quality Management District (SCAQMD), the California Energy Commission, the Port of Long Beach, and LA Metro, which were complemented by a settlement with a shipping company [11]. This model emphasises a collaborative, mission-driven model instead of a conventional, single-source government contract.

Table 5.2 compares the structures of these key ERS pilot projects, illustrating the different ways governments have partnered with the private sector to advance this technology.

5.4 Regulatory Frameworks and Standards

Although the funding and financing models cannot be ignored in the realisation of electric road system (ERS) projects, the existence of a strong and stable regulatory framework can be the prerequisite on which an interoperable, scalable, and safe network of ERS needs to be built. Single pilot projects can never be given ad hoc approvals to bring in the long-term, private investment that would see mass deployment. The detailed system would have to include a complex structure that could cover various levels, starting with high-level laws that would legalise the use of public right-of-way and ending with technical specifications of all aspects of the network. This may be modelled as a hierarchical interaction as Fig. 5.1 shows.

The scheme depicts the necessary flow of broad legislative power into a specific technical application. On top of that, a national or state parliament has enacted legislation (e.g., a National E-Mobility Infrastructure Act) to give ERS a legal framework and to be able to build it on the public roads, as well as to grant regulatory authorities. In the intermediary level, regulation, designed by the government agencies like the

Table 5.2 Comparison of PPP models in ERS pilot projects

Country/ project	Primary public partner(s)	Primary private partner(s)	PPP model/ structure	Funding structure	Key responsibilities (public vs. private)
Germany/ ELISA (eHighway)	Federal Ministry for Environment (BMU), Hessen Mobil (State Road Authority)	Siemens mobility	Government-funded R&D contract/ design-build	Public grants (e.g., federal funding for pilot projects)	**Public**: project management, funding, regulatory oversight. **Private**: technology provision, planning, construction, maintenance
Sweden/ Smartroad Gotland	Swedish Transport Administration (Trafikverket)	Electreon (leading a consortium)	Competitive tender for demonstration project	Public grants awarded through tender process	**Public**: funding, setting strategic goals, evaluation. **Private**: project leadership, technology deployment, testing, data collection
USA/ Carson, CA Pilot	South Coast Air Quality Management District (SCAQMD), CA Energy Commission, Port of Long Beach, LA Metro	Siemens Mobility, Mack Trucks, TransPower	Multi-agency public partnership for demonstration	Pooled public grants, legal settlement funds, private in-kind contribution	**Public**: co-funding, project oversight, environmental goals. **Private**: technology provision, vehicle development, system integration

Ministry of Transport or an energy regulator, makes a particular rule operational. This includes safety standards, grid-connection and land-use authorisation. On the lowest level, technical standards, which are normally drafted by national or international standards-development organisations (SDOs) like ISO, SAE or CENELEC in consultation with industry participants, specify the exact interoperability, performance and safety test requirements.

5.4.1 Safety and Electromagnetic Compliance (EMC)

The most prominent regulatory issue is safety. ERS brings in high-power electric infrastructure in the live and non-controlled state of the public road, which causes possible risks to be carefully controlled. Among the safety considerations to be made are:

- **Electrical Safety**: Electric shock safety to road users, maintenance personnel and first responders, especially during wet weather or during an accident.

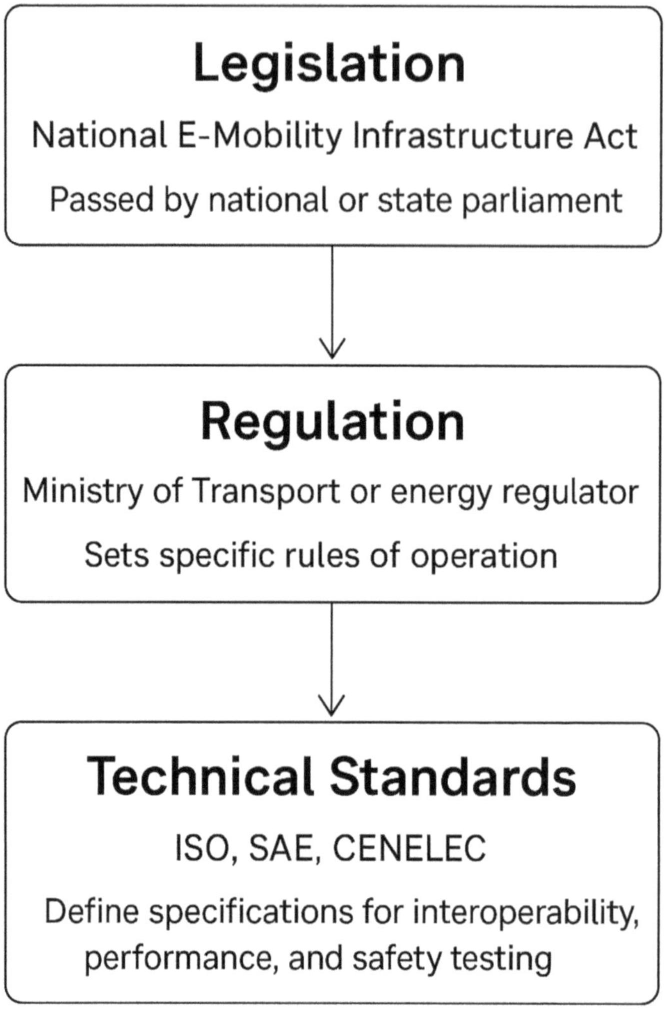

Fig. 5.1 The policy-technology interaction framework

- **Fire Safety**: Confidence in the integrity of the in-road/overhead infrastructure and the high-voltage systems in the vehicle to avoid fire, in particular, battery thermal runaway.
- **Mechanical Safety**: The physical strength of the infrastructure (e.g. overhead lines that can withstand high winds) and the vehicle interface (e.g. pantographs or receiver pads).

There are already safety standards of electric vehicles (EVs) that serve as a background. As an example, UNECE R100 requires very high safety standards of EV batteries, whereas ISO 6469 deals with the overall safety of electric road vehicles, such as electrical shock protection (UNCE R100; ISO 6469). Organisations like Underwriters Laboratories (UL), e.g. UL 2580 EV batteries and UL 2202 EV charging equipment, issue standards that provide specific testing guidelines for components. These standards were, however, mainly designed on a concept of static charging and might not be comprehensive enough to cover the peculiar risks of dynamic, in-motion power transfer.

In addition, electromagnetic compliance (EMC) is another important area of focus at the regulatory level. ERS systems should be designed such that they do not create electromagnetic interference with other vehicle electronics (e.g., ABS, airbags, communications systems), and external systems, e.g. radio communications and medical devices.

5.4.2 Interoperability and Standardisation

The requirement of interoperability is a prerequisite for ERS to be a feasible substitute for long-haul freight. A truck must have the capability of travelling through various territories, and possibly, countries, using ERS infrastructure offered by varying technology vendors. This also requires that the physical connection and the communication protocol between the road and vehicle should be standardised.

Standards of the communication aspect are the ISO 15118, which outlines the vehicle-to-grid (V2G) communication interface. This protocol allows smart charging, comprising authentication, billing, and load control between the EV and charger station. The ISO 15118, which is on wireless charging and pantograph charging, is thus directly relevant to ERS.

However, there is indeed a huge standards gap between the mature field of static charging and the emerging field of dynamic charging. SAE J2954 standard, as an example, is a large-scale industry-wide standard of wireless power transfer between stationary sources and light-duty vehicles. It sets standards in terms of interoperability, performance, safety and EMC, thus giving a clear direction that can be followed by manufacturers to produce compatible vehicles and chargers. Most importantly, the standard makes it clear that it is meant to be used with stationary applications and that the dynamic applications can be considered in the future. Such a gap is one of the main obstacles to scalability. Without a specific international standard in dynamic power transfer, there is a high probability of market fragmentation, a situation under which not only countries or regions use different technologies in a proprietary and incompatible way. This would create so-called technology islands and hamper the smooth cross-border freight activities, and drastically reduce the economic feasibility of a transnational ERS network.

5.4.3 *Road Access, Urban Planning, and Public Space*

The installation of ERS infrastructure has direct impacts on the right-of-way of the population, which means that the installation and operation of this infrastructure must be clearly regulated by legal means. This involves modification of current laws in relation to access to roads, installation of utilities and land use. An example is the municipal road access policies, which normally control any development in the right-of-way that belongs to the city and demand permits and agreements. Fire code regulations also specify a minimum unobstructed road width (e.g., 20 ft) that is needed to allow the passage of emergency vehicles, which can conflict with the location of ERS equipment or special lanes [12].

The urban and regional planning rules should be updated to include ERS. Such revisions are to include:

- **Zoning Ordinances**: Designating specific freight corridors or industrial zones as suitable for ERS deployment.
- **Long-Range Transportation Plans**: Integrating ERS corridors into metropolitan and statewide transportation plans to ensure coordinated development.
- **Permitting Processes**: Streamlining the permitting process for ERS installation to reduce project delays and costs.

The jurisdiction levels of this kind of infrastructure differ significantly in the regulatory environment. The European Union has traditionally embraced a top-down and harmonised method by imposing directives like the Alternative Fuels Infrastructure Regulation (AFIR) that incentivise standardisation across EU member countries. By contrast, the United States pursues a more scattered, bottom-up paradigm, and federal direction (e.g., of the FHWA) is associated with strong state-level control of transportation and energy regulation. Countries in Asia (such as China and South Korea) tend to adopt a government-led, technology-pull model, with national industrial policy and strategic investment in R&D being the main enablers. Such conflicting regulatory ideologies should have a far-reaching effect on ERS development rates and nature in both parts of the world.

5.5 Case Studies: National ERS Roadmaps

Strategies of developing electric road systems (ERS) have significant international differences and depend on differences in national priorities, industrial capacity, and policy incentives. As the analysis of the roadmap in Germany, Sweden, the United States, and South Korea reveals, ERS is not a homogenous notion but a versatile technology that can be used to solve a range of disparate core issues, including decarbonisation of the long-haul freight and urban air pollution, as well as technological innovation on display. This strategic intent variance explains the variance in technology choice, financing system and governance systems that each country uses.

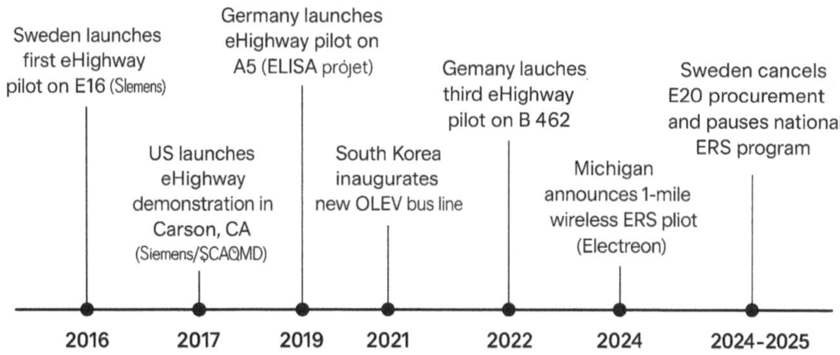

Fig. 5.2 Timeline of global ERS policy evolution (2015–2025)

Figure 5.2 shows an overview of the major policy and project milestones in the most successful countries.

A timeline chart would be presented here, showing key events such as:

- **2016**: Sweden launches first eHighway pilot on E16 (Siemens).
- **2017**: US launches eHighway demonstration in Carson, CA (Siemens/SCAQMD). Sweden launches eRoadArlanda pilot (Elways).
- **2019**: Germany launches eHighway pilot on A5 (ELISA project). South Korea inaugurates new OLEV bus line. Sweden launches Smart road Gotland pilot (Electreon).
- **2021**: Germany launches third eHighway pilot on B 462.
- **2022**: Michigan announces 1 mile wireless ERS pilot (Electreon).
- **2024–2025**: Sweden cancels E20 procurement and pauses national ERS program.

5.5.1 Germany

The ERS policy of Germany is clear-cut and targeted at decarbonisation of its massive heavy-duty freight industry. As trucks take up more than 70% of the volume of land-based transport and are a major source of emissions, ERS is considered an important tool to meet the national and EU climate objectives [13]. A joint initiative by the federal government has seen it spell out a roadmap which is centred on grand pilot projects on national roads.

The Federal Ministry for Digital and Transport (BMDV) leads the national strategy, and the Federal Ministry for the Environment, Nature Conservation, and Nuclear Safety (BMU) heavily funds the national strategy. This has seen the initiation of three key Overhead catenary technology eHighway pilot projects using Siemens overhead technology, including catenary technology [9]:

1. **ELISA Project (Hesse)**: A 10 km stretch on the A5 autobahn near Frankfurt, operational since 2019, connecting a logistics hub with the airport.

2. **FESH Project (Schleswig–Holstein)**: A section of the A1 autobahn near Lübeck, also launched in 2019, focusing on port-hinterland traffic.
3. **eWayBW Project (Baden-Württemberg)**: A 3.4 km route on the B 462 federal highway, launched in 2021, notable for testing the technology on a road with tighter curves and bridges [14].

Such projects are heavily funded; the BMU has allocated a total of €103 m to the three trials altogether, including the eWay BW project itself, which was allocated up to 26.8 m of federal funds of a total cost of 28 m for approximately three projects altogether [15]. Strong cooperation between the federal ministries and state road authorities (e.g., Hessen Mobil) and technology providers (Siemens) and logistics operators that operate the hybrid trucks are the governance model features of the hybrid trucks [16].

The scaling strategy is bold. An advisory body to the federal government, the National Platform for the Future of Mobility (NPM) has been proposing electrification of 300 km of autobahns by 2023 and 4,000 km by 2030 to cut greenhouse gas emissions in the transport sector by 40% [16].

5.5.2 Sweden

Sweden has also established itself as a world leader in ERS due to an ambitious climate agenda: a fossil-fuel-independent transport sector by 2030 and a net-zero economy by 2045, USAID [17]. The country undertook a technology-neutral research and demonstration programme beginning in 2013 under the auspices of the Swedish Transport Administration (Trafikverket) and is systematic in nature through all its activities [8]. Instead of pre-identifying a certain type of technology, Trafikverket employed pre-commercial procurement to fund and test several ideas in the real-life environment [18].

Key demonstration projects have included:

- **Overhead Catenary**: A 2 km stretch on the E16 motorway near Gävle, using Siemens eHighway technology, inaugurated in 2016 [19].
- **Conductive (In-Road)**: The eRoadArlanda project (2017–2019) tested a ground-level rail from Elways on a 2 km public road near Stockholm's Arlanda airport [19].
- **Inductive (Wireless)**: The Smart road Gotland project (2019–2023) demonstrated Electreon's wireless charging technology on a 1.65 km road, serving an electric bus and a 40-ton truck[10].

This step-by-step process was supposed to develop knowledge on a long-term ambition of an electrified national highway system. The initial permanent ERS was to be built by the government on a crucial section of the E20 corridor between Hallsberg and Örebro, which constitutes an important logistics route linking most of the Stockholm region and the Gothenburg region [20]. But this vision met with a great failure.

As of 2023, the E20 project was cancelled in the procurement process due to bids being higher than had been budgeted to [19]. A final report presented in December 2024 later advised against the creation of a national electric road network because it would not be cost-effective [15]. At the beginning of 2025, the Swedish government officially took off the project of the national infrastructure plan, which effectively stopped the large-scale ambitions of the country in regards to ERS [20]. This experience brings to emphasis the sheer economic and political difficulty with making the switch between successful, state-funded pilots, to one that is commercially viable and a large-scale network—a policy valley of death, where good technologies may fail unless a clear and inexpensive route to mass deployment exists.

5.5.3 United States

In America, the ERS is characterised by a disjointed, bottom-up type model, with federal research and development (R&D) and state pilot projects, which are motivated by regional demands. It does not have a single country ERS roadmap that is similar to the ones in Germany or Sweden.

At the federal level, the Department of Energy (DOE), through the Vehicle Technologies Office (VTO) is the key player, and it is providing funding to early-stage research into advanced methods of charging power, such as dynamic wireless power transfer (DWPT) [21]. This study, which is frequently performed in the National Laboratory like ORNL, aims at developing the technology behind in-road charging [22]. The infrastructure policy is administered by the Department of Transportation (DOT) and the Federal Highway Administration (FHWA). The national Electric Vehicle Infrastructure (NEVI) Formula Program, which is one of the provisions of the Bipartisan Infrastructure Law, allocates states up to 5 billion to develop a national system of fixed DC fast chargers at major highways [23]. At the moment, ERS projects are not directly evident as NEVI resources, which is another possible policy gap that may hinder alignment with the national EV charging plan [24].

State-level pilots have been the primary venues for ERS demonstration:

The main venues of ERS demonstration are state-level: In California, a one-mile pilot of the eHighway with Siemens catenary technology was a pilot in Carson (near the Ports of Los Angeles and Long Beach) beginning in 2017 [11]. The project was an alliance initiated by the South Coast Air Quality Management District (SCAQMD) and was majorly inspired by the intense necessity to mitigate serious air pollution caused by drayage trucks serving the ports [11].

5.5.4 South Korea

The ERS strategy of South Korea is structurally reinforced by the strong state support of technological innovation and its systematic integration into the national programs of smart-city development. Policy interest has mostly centred on the introduction and implementation of Online Electric Vehicle (OLEV) technology, a type of inductive (wireless) dynamic charging designed by the Korea Advanced Institute of Science and Technology (KAIST) [25].

The state has invested heavily in OLEV initiatives in the form of fiscal funds, with an early pilot project funded to the tune of about USD 69–78.5 million, setting the base scale of the first roll-out in municipal transport.

Introduction of OLEV into smart city systems is a central point of the general strategy. The efforts have been geared towards the utilisation of technology in order to come up with an efficient, sustainable, and intelligible urban transportation system, but such efforts have not been without challenges. Commercialisation of the OLEV has been difficult, although technological accomplishments and initial deployments have been realised. Several early bus routes were later cancelled due to the age of the infrastructure, and the ongoing state funding of the technology has been put under question and criticism. The larger policy of e-mobility in South Korea has remained strong, with large allocations of the technology to electric vehicle subsidies and charging capacity being managed by a triple ministry, including the Ministry of Trade, Industry and Energy (MOTIE), the Ministry of Environment (ME), and the Ministry of Land, Infrastructure and Transport (MOLIT) [26].

Table 5.3 provides a comparative summary of these four national approaches.

5.6 Policy Instruments and Economic Incentives

An effective business case of ERS needs a carefully developed package of economic incentives and financial policy instruments. They are also necessary as the cost difference between ERS-equipped vehicles and their traditional diesel versions is quite large, and addressing the significant upfront capital commitments related to the infrastructure development is necessary. The economic stimulus of ERS is different to the one used in passenger EVs; instead of acquiring subsidies in the form of a procurement solution, the key need of the ERS is the guarantee of high utilisation of the expensive road network. This therefore requires an effective policy to shift not only the emphasis on vehicle capital expenditure (CAPEX) but to incentivise operational expenditure (OPEX), thus making the use of electrified corridors possible as it is economically viable to the fleet operators.

A range of instruments can be deployed:

- **Direct Subsidies and Tax Credits**: They consist of the most widely used type of incentive in the wider EV market. In the case of ERS, it can include purchasing premiums on ERS-compatible trucks or tax credits for the fleet operators who

Table 5.3 Comparative analysis of national ERS roadmaps

Feature	Germany	Sweden	United States	South Korea
Lead government agency	BMDV, BMU	Trafikverket	DOE (R&D), state DOTs (pilots)	MOTIE, ME, MOLIT
Primary policy driver	Decarbonization of long-haul freight	Fossil-fuel independence, technology assessment	Air quality (CA), innovation (MI), federal R&D	Technology leadership, urban mobility, smart cities
Key technology tested	Overhead catenary	Tech-neutral (catenary, conductive, inductive)	Catenary (CA), INDUCTIVE (MI)	Inductive (OLEV)
Major pilot projects	eHighway (A1, A5, B 462)	eRoadArlanda, Smartroad Gotland, E16	Carson, CA; Detroit, MI	Gumi, Sejong OLEV buses
Funding model	Federal grants for large-scale pilots	Pre-commercial procurement, public grants	Pooled public grants (CA), state tender (MI), federal R&D grants	Direct government R&D and project funding
Scalability vision/status	Ambitious: 4,000 km by 2030 (NPM recommendation)	Vision for national network paused due to cost-effectiveness concerns	No national roadmap; state-led and project-specific	Commercialization challenges; focus on urban public transport

invest in the technology. Although Germany and Sweden are also offering a wide range of tax relief on zero-emission vehicles, their direct impact on the adoption of heavy-duty trucks might be relatively weakened compared to passenger vehicles because of the increased absolute capital costs of the latter. The FAME-II trend in India, which also includes incentives on the adoption of electric vehicles, is a model that could be modified or scaled to target the adoption of commercial vehicles using ERS in particular [27].

- **Infrastructure Financing and Grants**: This is a high capital need, and the direct public financing of infrastructure, therefore, becomes essential, especially during pilot and initial-scale stages. This is the same basic pattern in Germany, where the federal ministries are funding eHighway trails, and in Sweden, with Trafikverket funding demonstration projects, again, in the amount needed to demonstrate the scale of state spending on the large-scale deployment of infrastructure [28].
- **Operational Incentives**: Operational incentives are the most important in order to stimulate utilisation. These can involve high discounts or waived road charges on trucks using ERS corridors, which in turn will lead to a direct drop in the operating cost per-kilometre of use and provide a strong incentive to use electrified routes instead of more traditional ones. Another plan is to make the cost of electricity

supplied under ERS significantly less than the corresponding cost of diesel and, thus, create an apparent overall-cost-of-ownership benefit [29].

- **Innovative Financing Mechanisms**: With the maturity of the market, governments can use to access to the private capital by using innovative financing structures. Green bonds are a rather promising tool. These debt instruments would designate funds to projects with proven benefits to the environment, like clean transportation. ERS infrastructure projects, which have a direct impact on reducing the use of fossil fuels and greenhouse gases, would qualify well to be certified under initiatives such as the Climate Bonds Standard, therefore finding institutional investors to buy these assets with sustainability in mind [30].
- **Carbon Pricing and Credits**: Carbon taxes, or emissions trading schemes, which are market-based policies, raise the operating cost of diesel trucks, thus improving the relative economic appeal of ERS. In addition, ERS projects of a large scale, which can prove to reduce greenhouse gas emissions significantly, can issue carbon credits that will provide another source of revenue to project operators [31].

Table 5.4 provides a comparative overview of these economic instruments, their targets, and their respective advantages and disadvantages.

5.7 Institutional Coordination and Governance

The implementation of Electric Road Systems represents a complicated systems-integration issue far beyond the traditional boundaries of governmental ministries and government agencies. ERS is presented in the form of a special infrastructure, which physically and operationally integrates the road network with the electrical grid. A highway corridor fitted with ERS is a transport infrastructure element and at the same time a linear electrical substation. Its design is therefore based on both the freight traffic patterns (transport data) and local grid capacity (energy data); it is a kind of forced function that requires energy policy and transport policy to be deeply integrated.

Good governance of ERS thus requires formal cross-sectoral coordination mechanisms to be established. The institutional fragmentation, where siloed planning processes, divergent regulations, and lack of common data can be the most significant barrier to ERS, is often not addressed with technological or financial means but instead with institutional means: the Federal Ministry for the Environment financed the projects, and the Federal Ministry for Transport and its state-level agencies handled the work, displaying an effective, project-specific division of labour.

In the case of a national ERS program, the governance structure should be more permanent and empowered. This may be in the form of an inter-ministerial steering committee or special joint office. The development of the national ERS roadmap, harmonisation of regulations across the sectors, investment coordination and a unified contact with the partners of the private sector as well as other parties, would be the

Table 5.4 Comparison of economic incentive instruments for ERS

Instrument type	Target stakeholder	Country example(s)	Pros	Cons
Vehicle CAPEX subsidy	Fleet operators/ vehicle owners	Germany, Sweden (general EV incentives)	Lowers initial purchase barrier for ERS-compatible trucks	High public cost per vehicle; does not guarantee infrastructure utilization
Infrastructure grants	Road authorities/ tech providers/ CPOs	Germany, Sweden (pilot projects)	Enables initial deployment and de-risks private investment in unproven markets	High direct burden on public budgets; may not be sustainable for national scale-up
User toll discount/ exemption	Fleet operators/ vehicle owners	Norway (for general EVs)	Directly incentivizes use of the ERS corridor, driving up utilization and revenue	Reduces a potential revenue stream for infrastructure repayment; may require public subsidy to compensate operator
Favorable electricity tariffs	Fleet operators/ vehicle owners	(Proposed in studies)	Creates a strong and predictable TCO advantage over diesel	Requires regulatory intervention in electricity markets; may require subsidies if below market rate
Green bonds	Infrastructure operators/public agencies	(General transport projects)	Taps into large pools of private capital seeking sustainable investments; enhances project's green credentials	Requires robust certification and reporting; may not be accessible for entities with low creditworthiness
Carbon tax/ ETS	Fleet operators (via fuel cost)	European Union (ETS)	Market-based mechanism that automatically improves the competitiveness of low-carbon options like ERS	Politically sensitive; impact depends on the carbon price level

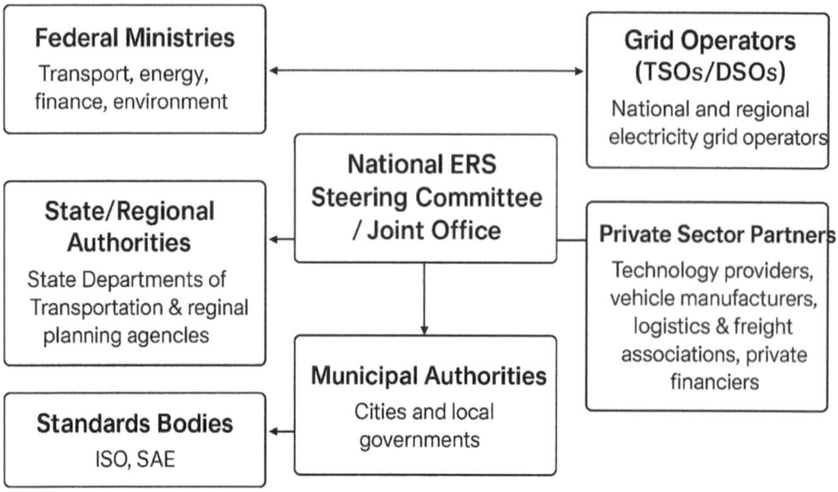

Fig. 5.3 A multi-stakeholder coordination framework for ERS governance

tasks of such a body. The history of Germany has created a Joint Agency for Electric Mobility (GGEMO) to foster its larger e-mobility efforts, so this body serves as a good precedent to such an institution [32].

This governance model must extend beyond government ministries to encompass a wide range of public and private stakeholders. A multi-stakeholder framework is essential for ensuring that the ERS ecosystem functions cohesively. Figure 5.3 illustrates a proposed coordination framework for ERS governance.

The proposed structure would have a National ERS Steering Committee or Joint Office at its centre. This executive branch would consist of senior officials of the main ministries and would have the mandate of providing strategic direction and policy coherence. It would directly coordinate and receive input from a broad range of stakeholders; these include:

- **Federal Ministries**: The Ministries of transport, energy, finance and environment offer the legislative and financial power.
- **State/Regional Authorities**: Implementation, permitting and integrating with local transport networks. It is the role of the State Departments of Transportation and regional planning agencies to do so on the ground.
- **Grid Operators (TSOs/DSOs)**: The electricity grid operators in the country and region are also important factors in determining grid capacity, network upgrades and technical standards of grid connection.
- **Private Sector Partners**: These comprise technology providers (e.g., Siemens, Electreon), vehicle manufacturers, logistics and freight associations and even the private financiers whose technical know-how and investment are critical.

- **Standards Bodies**: National and international standards organisations (e.g., ISO, SAE) must be engaged to develop the necessary technical standards for safety and interoperability.
- **Municipal Authorities**: National and international standards organisations (e.g., ISO, SAE) should be involved to come up with the required technical standards in terms of safety and interoperability.

Establishing such a formal governance mechanism is a very important initial move in any national ERS roadmap. It goes beyond the ad-hoc collaboration of pilot projects to create the institutional capacity needed to make long-term and systemic change. In fact, one of the most significant, but frequently ignored, advantages of undertaking ERS is the progress in these integrated planning models, which themselves are an effective resource in the handling of the more general and interrelated problems of the energy and transport transitions.

5.8 Policy Framework Proposal for India and Asia

Asian emerging economies, especially India, have a peculiar set of challenges and opportunities that make ERS an interesting strategy. Blistering economic development, rising urbanisation and over-dependency on road freight generate extreme stress on the transport industry, resulting in extreme congestion, air pollution, and reliance on imported fossil fuels. ERS in the case of India is not just an environmental policy, but it is a strategic instrument of pursuing fundamental national agendas of energy security, industrial growth through the program of making in India, and customer cost reduction through reduction of the logistics costs [33].

Based on the insights of the global case studies, the section outlines the recommendations of a five-pillar (plan) policy that can be applied to India with application to other developing Asian countries. This framework is based on the current policy environment in India, both its ambitious target of 30% EV penetration by 2030, the gigantic USD 1.4 trillion National Infrastructure Pipeline (NIP) and strategic initiatives like the Faster Adoption and Manufacturing of Electric Vehicles (FAME 2) scheme and the National Green Hydrogen Mission [27] (Fig. 5.4).

5.8.1 Pillar 1: Legislation and Standards Alignment

An unambiguous legal foundation is the first step.

- **National E-Freight Corridors Act**: India should enact dedicated legislation to provide the legal authority for ERS deployment on National Highways and other key routes. This act would empower central and state agencies to acquire right-of-way, set tariffs, and establish regulatory oversight.

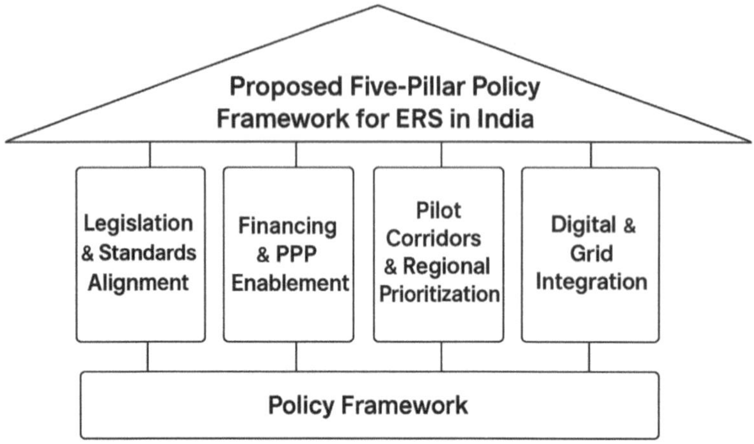

Fig. 5.4 Proposed five-pillar policy framework for ERS in India

- **Standards Harmonisation**: Records should be established under the Bureau of Indian Standards (BIS) with the involvement of automotive and power sector organisations, to form a national ERS standards committee. This committee would have the role of adopting and, in some cases, modifying international standards (e.g., ISO 15118 and future dynamic charging standards of SAE/ISO) towards interoperability and safety. This is in line with the assertion of NITI Aayog of providing a clear policy and mandates to drive the transition [34].

5.8.2 *Pillar 2: Financing and PPP Enablement*

The government has to develop a de-risked and conducive financial environment in order to draw the required capital.

- **NIP Integration**: India needs to establish specific laws to give the legal power in the implementation of ERS in National Highways and other major roads. This would enable the central and state agencies to purchase right-of-way, set tariffs and create regulatory control [35].
- **Blended Finance Facility**: Establish a special purpose fund as would be suggested by NITI Aayog on e-buses and e-trucks, combining public funds (budgets) and loans through multilateral development banks (e.g., World Bank, ADB), as well as through green bonds and other additional forms of private capital. This fund may offer low-interest loans or viability gap loans on ERS projects, rendering them bankable [34].
- **PPP Framework**: Standardise PPP contract templates to use in ERS projects (e.g. BOT or DBFO model) to simplify procurement and give clarity to the private investors in terms of risk distribution and revenue model.

5.8.3 *Pillar 3: Pilot Corridors and Regional Prioritisation*

By adhering to the strategy of NITI Aayog, a saturation rather than a thin distribution, the first steps should be taken on high-impact corridors in order to prove the feasibility and gain momentum [34].

- **Prioritise Key Freight Routes**: The first pilot initiatives should be implemented in the routes that have already been identified by the government to have with high volume of freight and strategic significance. These corridors, with major industrial and consumption centres, like Delhi–Jaipur, Bengaluru–Chennai, and Mumbai–Pune (as part of a larger Pune–Nashik corridor), are perfect candidates because the potential utilisation is high.
- **Phased Rollout**: Start with 50–100 km pilot zones on 2–3 of these routes. The lessons of these pilots, both in terms of cost and performance, as well as adoption by the user, ought to be used by a wider, gradual deployment program in the Golden Quadrilateral and other dedicated freight routes (Fig. 5.5).

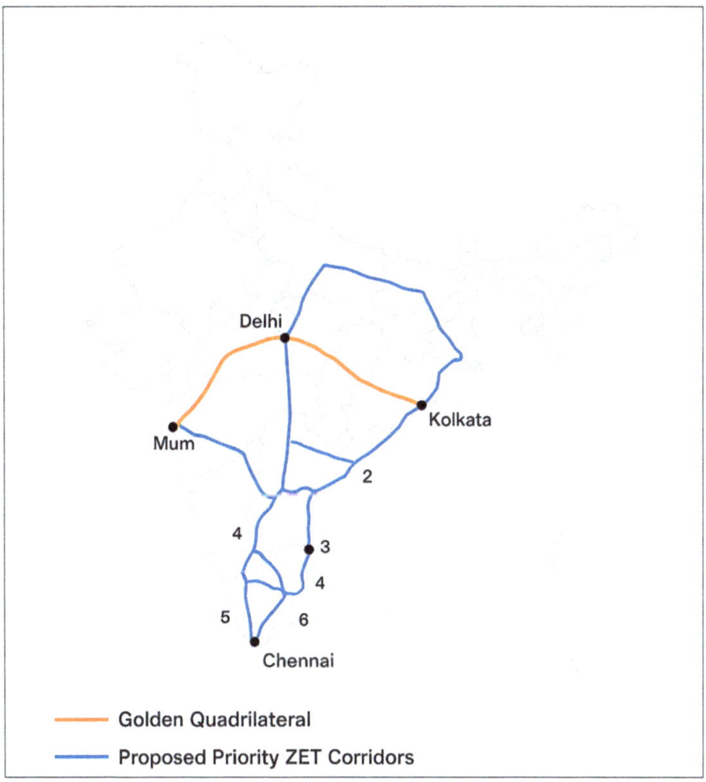

Fig. 5.5 Proposed priority ERS corridors for India

5.8.4 Pillar 4: Capacity Building and Skill Development

Deploying and maintaining a nationwide ERS network requires a skilled workforce that does not yet exist at scale.

- **Skill Development Mission**: Incorporate ERS-specific skills (ex: high power electrical engineering, pantograph maintenance, grid management software) into programmes at the national level, like the Skill India Mission.
- **Academic-Industry Partnerships**: Encourage a relationship between academic institutions, such as the Indian Institutes of Technology (IITs) that are already involved with ZET research and the technology suppliers themselves [36]. This can result in specialised curricula, training facilities and a pool of qualified engineers and technicians.

5.8.5 Pillar 5: Digital and Grid Integration

ERS will have to be intelligent and grid-conscious at the very beginning:

- **Smart Grid Mandates**: Requirements that make all ERS infrastructure smartly charged, so that dynamic load control can be used according to grid conditions. This is essential in incorporating large and unpredictable loads without destabilising the electric grid [37].
- **Battery Management and Leasing**: Adopt the idea of NITI Aayog of having a battery passport system [34]. This electronic account of battery history and health would be the key to allowing a battery leasing business. The initial capital expenditure of an ERS truck can also be dramatically lowered by letting the fleet operators lease the batches, rather than owning them, and thus speeding up adoption.

India can develop a complete and solid ecosystem of ERS by designing its approach around these five pillars, which can make a technological possibility a strategic national asset.

5.9 Future Outlook

The success and global effectiveness of Electric Road Systems in the long term will rely on how the process of changing the collection of localised, national pilot initiatives into an aligned, compatible, and universally collaborative ecosystem is made. The three major trends in the future of ERS governance will be global harmonisation of standards, improved cooperation at the international level, and the introduction of flexible governance structures that are able to cope with the technology.

Global Harmonisation of Standards: The Standards chasm between the dynamic and the static charging is one of the most significant impediments to cross-border deployment of ERS [38]. The future of ERS in international freight transport depends on the creation and implementation of a common standard of dynamic transfer of power. Such concerted research efforts as the Swedish Swedish-German–French CollERS project represent a key model to the process and entail pulling together national transport authorities as well as researchers to agree on technical specifications and policy direction towards a global standard needed to avoid technology lock and developing a genuinely global market in which ERS vehicles and infrastructure can operate.

International Cooperation: With the growing interest of countries to learn ERS, the importance of knowledge and best practices sharing increases. The global arena can also hasten the learning curve of the emerging economies and avoid unnecessarily expensive errors. Technology transfer and co-development of policies can be conducted through dialogues between existing and future ERS leaders, like possible EU-India partnerships of sustainable transport. Also, the regional organisations, including the United Nations Economic and Social Commission of Asia and the Pacific (UNESCAP), should be central in supporting ERS as a part of larger sustainable transport agendas in the Asian region to enable the member states to formulate their national roadmap [39].

Adaptive Governance: ERS is a new technology in the fast-growing energy and transport environment. The policy system adopted today is rigid and may become stagnant in ten years. Hence, governments need to embrace an adaptive mode of governance [34]. This is establishing regulatory systems that are adaptable and learning oriented. The policies must be designed in such a way that they can be regularly reviewed and amended in response to the data and experience obtained during the ongoing pilots and preliminary deployment. This cyclic approach will make sure that regulations do not get behind the curve of technological changes and market conditions to encourage constant innovation instead of suppressing it [36].

Finally, the policy drive towards ERS resonates well with the sustainable development agenda of the world. ERS is a nexus technology that can support several UN Sustainable Development Goals (SDGs), which makes it particularly strong as a policy case to be a strategic investment in national governments.

- **SDG 7 (Affordable and Clean Energy)**: The creation of an enormous, predictable electricity demand along transport routes enables ERS to become an anchor load of new renewable energy generation projects, enabling them to be more readily integrated into the grid [37].
- **SDG 9 (Industry, Innovation, and Infrastructure)**: ERS is a new paradigm of sustainable and resilient infrastructure, whereby power electronics, vehicle production, and smart grid control are being innovated [40].
- **SDG 11 (Sustainable Cities and Communities)**: ERS can ensure that air and noise pollution in cities is drastically cut as a result of urban freight and public transport electrification, making urban living places healthier and lovable [40].

- **SDG 13 (Climate Action)**: ERS presents a direct and scalable opportunity to decarbonise the hard-to-abate heavy-duty transport sector, which is a key element in meeting national and global climatic goals [41].

Governments can construct more effective political coalitions and open up a variety of sources of sustainable financing to push their development by positioning ERS not as a simple transport project, but as a strategic facilitator of more general SDG commitments.

5.10 Conclusion

The experiences of the most innovative countries have offered relevant information in terms of policy. A case study of Germany, Sweden, the United States and South Korea shows that there is no universal solution since national roadmaps are influenced by a variety of strategic drivers, which may include decarbonisation of long-haul freight, development of breakthrough technology or reduction of air pollution in cities. The identified policy mismatch, i.e., the inability of the policy of consumer-oriented incentives on electric vehicles to correspond to the B2B infrastructure development that Electric Road Systems (ERS) needs, has become a pressing issue. This observation highlights the need to embrace the idea of de-risking large-scale capital projects, and more importantly, to encourage the use of infrastructure after construction.

Public–Private Partnerships are still necessary, but they need to be redefined in the form of Innovation Partnerships where they realise the mutual risk of commercialising nascent technologies. In addition, the gap between the paradigms of the fixed and dynamic charging, that is known as the standards chasm, has to be overcome by means of international cooperation to ensure the interoperability required to achieve a seamless cross-border network. And, arguably, most fundamentally, the nature of road and grid infrastructure embedded within ERS is an obligatory device of breaking down institutional silos, which requires unprecedented communication between ministries of transport, energy, and finance.

Public–Private Partnerships are essential, but must be framed as "Innovation Partnerships" that acknowledge the shared risk of commercialising a new technology. Furthermore, the "standards chasm" between static and dynamic charging must be bridged through international collaboration to ensure the interoperability required for a seamless, cross-border network. Perhaps most importantly, the inherent fusion of road and grid infrastructure in ERS acts as a "forced function" for breaking down institutional silos, demanding unprecedented coordination between transport, energy, and finance ministries.

For emerging economies like India, ERS offers a multi-faceted strategic opportunity to advance energy security, industrial development, and economic competitiveness. The proposed five-pillar framework—encompassing legislation, finance, prioritised pilots, capacity building, and grid integration—provides a tailored roadmap for navigating this complex landscape.

In the case of emerging economies like India, ERS has a complex strategic business opportunity to increase energy security, industrial growth, and economic competitiveness. A five-pillar framework, including legislative reform, financial mechanisms, prioritised pilots, capacity building and grid integration, provides a specific roadmap to navigate in this complex landscape.

References

1. Siemens Mobility. eHighway—dynamic charging infrastructure. https://www.mobility.siemens.com/global/en/portfolio/road/ehighway.html. Accessed 10 Oct 2025
2. PIARC. Electric road systems: a solution for the future. https://www.piarc.org/en/order-library/29690-en-Electric%20road%20systems:%20a%20solution%20for%20the%20future. Accessed 10 Oct 2025
3. Dataintelo. Electric road system market research report 2033. https://dataintelo.com/report/electric-road-system-market. Accessed 10 Oct 2025
4. United Nations Economic Commission for Europe (UNECE). Electric mobility. https://unece.org/sustainable-energy/electric-mobility. Accessed 10 Oct 2025
5. International Energy Agency (IEA). Policies to promote electric vehicle deployment—global EV outlook 2021. https://www.iea.org/reports/global-ev-outlook-2021/policies-to-promote-electric-vehicle-deployment. Accessed 10 Oct 2025
6. International Transport Forum (ITF–OECD). Transport–energy nexus to deliver climate goals and a just transition. https://www.itf-oecd.org/sites/default/files/summary_-_transport_energy_ministers_roundtable.pdf. Accessed 10 Oct 2025
7. Highways Today. The ultimate guide to public–private partnership financing. https://highways.today/2025/04/13/public-private-partnership-financing/. Accessed 10 Oct 2025
8. McKinsey & Company. Electric road systems. https://www.mckinsey.com/industries/infrastructure/global-infrastructure-initiative/site-visits/electric-road-systems. Accessed 10 Oct 2025
9. Siemens. Test field eHighway—project eWayBW. https://press.siemens.com/global/en/pressrelease/test-field-ehighway-project-ewaybw. Accessed 10 Oct 2025
10. Electreon. Gotland, Sweden electric road case study. https://electreon.com/projects/gotland. Accessed 10 Oct 2025
11. Siemens USA. First U.S. eHighway demonstration running in California. https://www.siemens.com/us/en/company/press/press-releases/mobility/first-us-ehighway-demonstration-running-in-california.html. Accessed 10 Oct 2025
12. National Association of City Transportation Officials (NACTO). Best practices—emergency access in healthy streets. https://nacto.org/wp-content/uploads/Best-Practices-Emergency-Access-in-Healthy-Streets.pdf. Accessed 10 Oct 2025
13. eHighway.SH. Frequently asked questions. https://ehighway-sh.de/en/faq/. Accessed 10 Oct 2025
14. OEM Off-Highway. Siemens begins third eHighway trial in Germany. https://www.oemoffhighway.com/market-analysis/trends/press-release/21551491/siemens-industry-inc-siemens-begins-third-ehighway-trial-in-germany. Accessed 10 Oct 2025
15. Bureau of Energy Efficiency (BEE), Government of India. Energy efficiency and conservation programmes for electric mobility. https://beeindia.gov.in. Accessed 10 Oct 2025
16. NITI Aayog. Battery swapping policy draft 2022. https://www.niti.gov.in/sites/default/files/2022-04/Battery_Swapping_Policy_Draft.pdf. Accessed 10 Oct 2025
17. Department of Heavy Industries (DHI), Government of India. FAME India scheme—phase II. https://dhi.nic.in/UserView/index?mid=1388. Accessed 10 Oct 2025

18. NITI Aayog and Rocky Mountain Institute. India's electric mobility transformation: progress to 2030. https://rmi.org/insight/indias-electric-mobility-transformation. Accessed 10 Oct 2025
19. Central Electricity Authority (CEA). National electricity plan 2023—transmission and distribution infrastructure. https://cea.nic.in. Accessed 10 Oct 2025
20. World Economic Forum (WEF). Decarbonizing road transport 2023. https://www.weforum.org/reports/decarbonizing-road-transport-2023. Accessed 10 Oct 2025
21. Asian Development Bank (ADB). Electric vehicle ecosystem development in Asia. https://www.adb.org/publications/electric-vehicle-ecosystem-asia. Accessed 10 Oct 2025
22. World Bank. Financing electric mobility transitions in developing economies. https://www.worldbank.org/en/topic/transport/publication/electric-mobility-financing. Accessed 10 Oct 2025
23. Indian Institute of Technology Madras (IITM). Centre for battery engineering and electric vehicles (CBEEV)—dynamic charging research. https://cbeev.iitm.ac.in. Accessed 10 Oct 2025
24. World Resources Institute (WRI India). Roadmap for electric mobility in Indian cities. https://wri-india.org/publication/roadmap-electric-mobility-india. Accessed 10 Oct 2025
25. eRoadArlanda. Electric road pilot project in Sweden. https://eroadarlanda.com. Accessed 10 Oct 2025
26. Japan Ministry of Land, Infrastructure, Transport and Tourism (MLIT). Smart mobility and electrification initiatives. https://www.mlit.go.jp/en/. Accessed 10 Oct 2025
27. Economic Times Auto. India plans pilot electric highway between Delhi and Jaipur. https://auto.economictimes.indiatimes.com/news/india-electric-highway-delhi-jaipur/. Accessed 10 Oct 2025
28. International Energy Agency (IEA). Executive summary—grid integration of electric vehicles. https://www.iea.org/reports/grid-integration-of-electric-vehicles/executive-summary. Accessed 10 Oct 2025
29. The Hindu BusinessLine. India's electric highways: next frontier for green logistics. https://www.thehindubusinessline.com/economy/india-electric-highways-2025/. Accessed 10 Oct 2025
30. Frontiers. Impacts of electric road systems on the German and European transport network. https://doi.org/10.3389/fenrg.2021.631200/full. Accessed 10 Oct 2025
31. NITI Aayog. Promoting clean energy usage through accelerated localization of E-mobility value chain. https://www.niti.gov.in/sites/default/files/2023-07/Niti-Aayog_Report-VS_compressed_compressed.pdf. Accessed 10 Oct 2025
32. Education Post. Govt picks 10 freight corridors for electric and hydrogen truck deployment. https://educationpost.in/news/environment/govt-picks-10-freight-corridors-for-electric-and-hydrogen-truck-deployment. Accessed 10 Oct 2025
33. EFAST India. Powering India's electric truck corridors through strategic infrastructure development. https://efastindia.org/powering-indias-electric-truck-corridors-through-strategic-infrastructure-development. Accessed 10 Oct 2025
34. ResearchGate. Governance frameworks for renewable energy development, energy transition, and public governance. https://www.researchgate.net/publication/385769730_Governance_Frameworks_for_Renewable_Energy_Development_Energy_Transition_and_Public_Governance. Accessed 10 Oct 2025
35. NITI Aayog. Electric vehicle charging infrastructure implementation. https://www.niti.gov.in/sites/default/files/2023-02/EV_Handbook_Final_14Oct.pdf. Accessed 10 Oct 2025
36. Frontiers. A governance framework for implementation of scientific and engineering innovation in buried infrastructure systems. https://doi.org/10.3389/frsc.2022.765577/full. Accessed 10 Oct 2025
37. ESMAP. The outlook for SDG 7. https://trackingsdg7.esmap.org/data/files/download-documents/SDG7%20-%20Report%202024%20-%20Chapter6-TheOutlookForSDG7.pdf. Accessed 10 Oct 2025

38. SAE International. Wireless power transfer for light-duty plug-in/electric vehicles and alignment methodology, J2954_202010. https://www.sae.org/standards/j2954_202010-wireless-power-transfer-light-duty-plug-electric-vehicles-alignment-methodology. Accessed 10 Oct 2025
39. UN Department of Economic and Social Affairs. Goal 7. https://sdgs.un.org/goals/goal7. Accessed 10 Oct 2025
40. UN Global Compact. SDG 07—SDG blueprint. https://blueprint.unglobalcompact.org/sdgs/sdg07/. Accessed 10 Oct 2025
41. United Nations. Energy action—sustainable development goals. https://sdgs.un.org/sites/default/files/2023-01/SDG%207%20Policy%20Briefs%202021.pdf. Accessed 10 Oct 2025

Chapter 6
ERS for Emerging Asia: Infrastructure Readiness and Challenges

6.1 Introduction

The effective implementation of Electric Road Systems (ERS) in economies in their infancy is not only a technological issue but also a basic reliance on the existing state of the infrastructure base of a country. ERS is an overlay technology; its performance, reliability and economic viability are directly limited by the quality of the substrate upon which it is implemented, namely, the road networks and the power grid. The decarbonisation of heavy freight transport is a promise that is already promising, but the infrastructural realities that lie on the way to its implementation differ considerably within the Asian continent. This chapter analyses the preparedness of four major emerging Asian economies of India, Bangladesh, Thailand, and Indonesia in regards to their ability to sustain such a transformative system that transcends a mere inventory of assets to the more subtle nature of the infrastructure that underpins them.

This assessment is structured around a tripartite definition of "readiness," which forms the analytical framework for the subsequent sections:

1. **Engineering Readiness**: This term is used to refer to the physical ability of road networks to carry ERS hardware. It includes key parameters of civil-engineering strength (pavement), stability of the subgrade, uniformity of the surface, and sufficiency of drainage, and all of these are critical to the long-term integrity of embedded or overhead ERS elements.
2. **Grid Readiness**: This evaluates the capability of national and regional electrical grids to supply stable, reliable, and sufficient power to dynamic, high-load ERS segments. It considers not just raw generation capacity but also the robustness of transmission and distribution (T&D) networks, grid stability under fluctuating loads, and the potential for integration with renewable energy sources.
3. **Institutional Readiness**: This dimension evaluates the enabling environment, which is not physical (that is, the existence of supportive policy frameworks

R. Agnihotri, *Electrified Road Systems (ERS) for Emerging Asia*, Springer Tracts on Transportation and Traffic 24, https://doi.org/10.1007/978-3-032-16927-3_6

to electric mobility, systems to support robust financing, and simplified regulatory and land-acquisition procedures that can support complex, multi-sectoral infrastructure projects).

The Asian emerging world of transport and energy is highly heterogeneous. Countries are in different levels of economic development, and their levels of infrastructure maturity, investment capacity, and institutional depth vary; a universal solution to ERS implementation is thus unsustainable. The chapter uses a comparative approach, based on verifiable statistics on these sources, including the Asian Development Bank (ADB), the World Bank, and country ministries, to build a granular country-specific image of readiness.

The first step in the analysis will be setting a quantitative baseline of the regional road and grid infrastructure. It will then perform a technical examination on national standards of civil engineering and road quality indicators. Freight-logistics trends will thereafter be analysed to determine the most cost-effective corridors to use E-road. Such a multi-dimensional evaluation will then lead to the creation of a new ERS Readiness Index (ERI), a composite measure that will be used to rank the four countries in the main dimensions of readiness. In the end, the chapter shall come up with the greatest issues and bottlenecks, namely, technical, economic, institutional, and social, followed by strategic and country-specific interventions and pathways of adapting to the change in phases. With this overarching preparation estimation, the detailed cost–benefit analysis and funding model in Chap. 7 will have the required background.

6.2 Regional Overview of Road Infrastructure

The macro-level analysis of the current road networks of India, Bangladesh, Thailand and Indonesia indicates the magnitude of the opportunity as well as the size of the challenge in the implementation of ERS. Although the total network lengths seem to be very large, the type of pavement and lane structure, as well as administrative classification, should be analysed more closely to differentiate between the total network and the effective network that should be introduced into the first version of ERS.

The largest road network in the world belongs to India, with over 6.612 million km [1], but only 146,145 km of this national road network are of the National Highways (NH) size that can accommodate modern transport systems, with only 321 of that entire system operating as four-lane or wider [2]. This discrepancy explains why the ERS challenge in India is not the volume of roads, but rather the volume of high-quality, high-capacity corridors.

The total road network of Bangladesh consists of around 375,000 km of road, which is a large proportion of the land area, although only a small percentage of this is paved roads, so this limits the usage by any technology that demands high-quality surface infrastructure [3].

Thailand has an extensively developed and large road network, estimating that an estimated total system length of between 390,000 and 700,000 km, covering national and major routes, as well as rural and local routes [4]. The Thai road location is good in terms of the availability of basic roads.

Since Indonesia is a large archipelago, road transport plays an important role, with a total length of 548,366 km of the network as of 2020 [5]. The country is largely concentrated on the island of Java, which is the economic and demographic centre of the country, and considerable development is also performed on the island of Sumatra.

This briefing exercise has shown that the crude measure of total road network is fallacious to evaluate ERS preparedness. Most of the roads in the countries are of rural and local nature or of poor quality, that is not fit to be used in the initial deployment of ERS. The key measure is the length of the effective network—the multi-lane, paved, and well-maintained national highways, expressways, or economic corridors that support most of the long-haul freight traffic. It is not the electrification of a complete national network, then, of which strategic challenge ERS adoption must be sought, but the need to identify and upgrade a few thousand kilometres of these vital economic arteries (Table 6.1).

The Asian Highway (AH) network, a project coordinated by UNESCAP to improve connectivity across the continent, provides a useful framework for identifying corridors with the highest potential for ERS [14]. Key routes such as AH1 and AH2 traverse India, Bangladesh, and Thailand, while AH41 is a critical corridor within Bangladesh [13]. These routes often overlap with national economic corridors, such as India's Golden Quadrilateral, the Dhaka-Chattogram axis in Bangladesh, and major arteries connecting to Thailand's Eastern Economic Corridor. As shown conceptually in Fig. 6.1, these corridors represent the most logical starting points

Table 6.1 Comparative road statistics (2023/latest available)

Metric	India	Bangladesh	Thailand	Indonesia
Total road network (km)	6,617,100 [1]	375,000 [6]	~702,989 [7]	548,097 [8]
National/state highway length (km)	146,145 (NH) [9]	22,719 (RHD) [10]	52,000 (DOH) [11]	47,817 (State) [8]
Expressway length (km)	~3,900	~132 (Operational) [6]	~200	3,000 [12]
Paved road share (%)	~65%	~29% [13]	~98%	~58%
Road density (km per 1,000 km^2)	2,012	2,527	1,371	287

Note Data compiled from multiple sources with varying years and methodologies; figures are indicative. Paved road share and density are calculated based on a mix of reported data and estimates

Fig. 6.1 Map—Asian highway corridors with ERS potential

for pilot ERS projects due to their high traffic volumes and strategic economic importance.

6.3 Civil Engineering and Construction Standards

The technical practicability of the installation of electric road systems (ERS) relies at a fundamental level on the compatibility between the national civil engineering standards and the road construction standards, and the specific demands dictated by ERS hardware. The demands on the ERS technologies are not limited to those of conventional traffic loads but may be conductive rails embedded in pavement, inductive coils embedded in the surface or overhead catenary systems. In this section, a survey of the existing standards in the four countries considered in the context of ERS incorporation will be conducted to analyse their appropriateness in the implementation of ERS.

The road building in India is controlled by the complete codes issued by the Indian Roads Congress (IRC). In the case of flexible pavements (which form the bulk of the network), such as IRC:39-1986 gives specific guidelines describing how to construct layers based on traffic assessment (measured in million standard axles) and subgrade strength, typically expressed in the use of the California Bearing Ratio (CBR) [15]. Although the codes are robust to conventional loads, they do not include guidelines related to the integration of electrical components, such as the definition of non-metallic aggregates in top-most pavement layers, inductive ERS, or structural considerations.

The Department of Highways (DOH) of Thailand uses strict criteria, especially in its modern intercity motorway [16]. The pavement structure of the Bang Pa-in-Nakhon Ratchasima motorway has a pavement structure of 28 cm designed to ensure high-speed and heavy vehicles can use the road safely and comfortably [17]. The focus of the Department of Highways (DOH) in establishing a safe and convenient highway system gives the implication that the quality of a surface and its integrity are of high value, which is beneficial to ERS. But, just like IRC standards, these standards are traffic-oriented and do not include any particular instructions on how to enclose or mount electrical equipment.

Indonesia has a system of national standards, the Directorate General of Highways, Bina Marga, which has a Norms, Standards, Procedures, and Criteria (NSPK), which design of layers-thickness is based on traffic load and sub-ground CBR value [18]. This is a change to the older 2002 method, which uses the subgrade resilient modulus [18].

The use of the concept of flexible pavement is mostly appropriate to the climatic conditions of Bangladesh, and the standards of the pavement design are mostly based on the projected number of Equivalent Standard Axles (ESAs), which is 8.2 tons, in the lifespan of the road, rather than the integration of the electrical systems [19].

A comparative study of these national standards shows that there is a vital gap. Although the four countries have mature systems of developing durable roads, none of them has integrated the needs of ERS. Inductive charging systems, as an example, need a tight burial depth and a sensitive choice of pavement material to eliminate electromagnetic interference; conductive systems need an exceptionally smooth surface and advanced drainage to eliminate water soaking and short circuiting. Catenary systems require stable shoulders and undercarriage that will be able to support masts at periodic intervals. Lack of such specifications in the existing national codes means that any ERS project will require custom-designed engineering designs and approvals, thus posing a big technical and regulatory challenge. This demonstrates an institutional capacity gap: the necessity to work out new technical standards or to officially introduce and adapt international ones (e.g. European ERS projects) as a precondition of large-scale implementation (Table 6.2).

6.4 Electrical and Grid Infrastructure Readiness

The renewable energy of an electric road system is directly associated with the capacity, stability, and reliability of the national power grid that has to supply it. An ERS presents a special form of electric load, mobile, highly fluctuating and consuming a large amount of power at particular locations along a corridor. This part assesses the grid preparedness of the four countries not only in terms of the total generation capacity, but also on the key issues of transmission, distribution, and the quality of the system as a whole.

Table 6.2 Compatibility of national road standards with ERS requirements

ERS requirement	India (IRC)	Bangladesh (RHD/AASHTO)	Thailand (DOH)	Indonesia (Bina Marga)
Subgrade strength (CBR/ modulus)	Sufficiently specified	Sufficiently specified	Sufficiently specified	Sufficiently specified
Pavement thickness for heavy loads	Sufficiently specified	Sufficiently specified	Sufficiently specified	Sufficiently specified
Surface uniformity (IRI tolerance)	Partially specified (in quality control)	Partially specified (in quality control)	Partially specified (in quality control)	Partially specified (in quality control)
Drainage capacity	Specified for road integrity, not for electrical safety	Specified for road integrity, not for electrical safety	Specified for road integrity, not for electrical safety	Specified for road integrity, not for electrical safety
Non-metallic aggregate specs	Not specified	Not specified	Not specified	Not specified
Shoulder stability for masts	Partially specified (for signage/ lighting)	Partially specified (for signage/ lighting)	Partially specified (for signage/ lighting)	Partially specified (for signage/ lighting)

India boasts a massive installed generation capacity of 457 GW as of late 2024, with a large and ambitious National Electricity Plan targeting 458 GW of peak demand by 2032 [20]. The country has reached almost complete access to electricity, but consumption per person is relatively low at 1,395 kWh (2023–24), and the latent demand is significant [20]. The biggest barrier has been large transmission and distribution (T&D) losses, but these are declining steadily [21].

Bangladesh is a contradictory example. The overall generation capacity of the country amounted to 30.7 GW [22]. Despite the abundance of surplus, the grid is not very reliable, with constant power disruptions that would make the system unusable to a dynamic and critical load like ERS.

The grid of Thailand is more stable and reliable due to the cumulative installed capacity of 56.8 GW in 2023, which is relatively high in terms of electrification rates and T&D network control (power grid distribution). The grid is, nevertheless, very reliant on natural gas generation (almost 59%) [23]. The Power Development Plan (PDP) of the government, which projects the addition of 28 GW of renewable capacity by 2037, is a huge prospect to energise future ERS corridors with clean energy and support the development of infrastructure in line with national climate objectives.

The national electricity grid of Indonesia, which contains an installed capacity of about 73 gigawatts as of the end of 2023, is inherently fragmented because of the archipelagic form of the nation [24]. JavaBali subgrid is the most developed and strong part serving the major economic centres. The cumulative distance of the transmission network is 87,691 km, and the overall energy losses are only 8.57% in 2023 [24]. A key issue of Indonesia is the grid stability and capacity of the islands other than Java, whereby, in the future, the implementation of an Electrified Road System (ERS) could be considered.

This discussion shows that raw generation capacity is not a good measure of ERS preparedness. It depends on the state of the transmission and distribution (T&D) network. Even a system that produces excess, as the case of Bangladesh does, will not be ready to operate according to ERS when compared to the Thai system, which might have a relatively smaller amount of surplus but would be more stable and reliable. Implementing an ERS would place intense pressure on the local distribution networks, and upgrading substations and medium-voltage feeders (e.g. 1133NV) along the chosen corridors would be necessary. The main agenda in the ERS enablement, therefore, needs to be on T&D modernisation as opposed to generation. The conceptual idea developed in Fig. 6.2 illustrates how ERS corridors can be strategically positioned with current grid infrastructure to reduce the connection costs (Table 6.3).

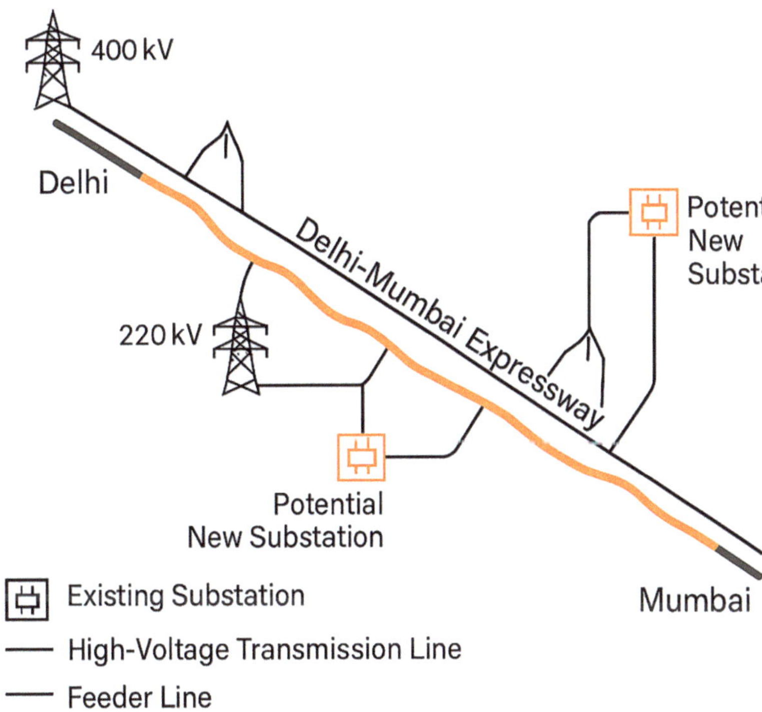

Fig. 6.2 Grid-highway overlay diagram (example: Delhi-Mumbai Corridor, India)

Table 6.3 Grid readiness parameters by country (2023/latest available)

Metric	India	Bangladesh	Thailand	Indonesia
Total installed capacity (GW)	457	30.7	56.8	73.0
Access to electricity (%)	100	93	~100	98.3
T&D losses (%)	~16% (est. 2022)	7.8 (2022)	~6.5% (est. 2022)	8.57
Per capita consumption (kWh)	1,395 (2023–24)	574 (2022)	~2,900 (est. 2022)	~1,173 (est. 2022)
Share of renewables (incl. hydro) (%)	43.0	~4.5	~13.5	~13.1 (PLN group)

Note Data compiled from multiple sources with varying years. T&D losses and per capita consumption for some countries are estimates based on the latest available World Bank/IEA data trends

6.5 Freight Logistics and Traffic Patterns

The strategic consideration of implementing an ERS is based on identifying areas that have the highest concentration of freight traffic and, hence, making sure that the large capital outlay is translated to the highest economic and environmental returns. Freight logistics analysis and traffic patterns analysis, consequently, cannot be omitted in coordinating the future ERS routes with the economic geography of every country.

Road transport in India is the unquestioned workhorse of the logistics sector, with freight volumes in these routes accounting for 71% of the total modal share [25]. Traffic volumes on these arteries are extremely high and culminate in high congestion [2]. This higher truck density makes these corridors an ideal candidate for ERS. However, India is also simultaneously undertaking a titanic project in the form of rail-based Dedicated Freight Corridors (DFCs) to replace road transport with rail transport [26]. This concomitant development poses a hugely important strategic issue since ERS-powered trucking would either compete with the new, efficient freight trains over long distances or become a complementary service, offering electrified first- and last-mile connectivity to and from DFC terminals and multimodal logistics parks. An ERS strategy that works successfully in India will probably have to be symbiotic and not antagonistic towards the growing DFC network.

The economy of Bangladesh is very much dependent on a few major arteries, with road freight occupation for 69.4% of the logistics market [27]. The road transport system is the economic backbone of Bangladesh, where the Dhaka-Chattogram highway (part of AH 41) is the one that carries the majority of the imports and exports between the two regions, and the capital and the main seaport. This route experiences chronic overpacking; hence, the logistics solutions, including ERS, are a priority to enhance its capacity, such as the ERS [27], as recently the Padma Bridge was opened and has provided a viable southwest corridor connecting Dhaka to the Benapole land port and the Mongla seaport as a secondary axis where the necessary ERS can be deployed.

Road freight is the dominant mode of transport in Thailand, and it is reported that there were 484.8 million tonnes in 2016 [28]. Freight movement with the EEC, located in the hub of the Bangkok Metropolitan Region and the industrial powerhouse of the Eastern Economic Corridor (EEC), which connects Bangkok to the deep-sea port of Laem. The major goods that are transported along the road are the industrial goods, agricultural commodities like sugar cane, and construction materials [28].

The road freight market in Indonesia has a freight market of more than USD 53 billion, with the long-haul services in Java being the most important, with almost three-quarters of the market worth being directly on the main economic flows in the country [29]. The reason is that these toll roads, with their controlled access and improved-quality constructions, are the natural and most promising candidates to be deployed in ERS in that they literally follow the main economic flows in the country.

6.6 Road Quality and Safety Indicators

In addition to the macro-level presence of road networks, the micro-level state of the surface pavement is a key technical factor of ERS success. The road quality, which has been described as roughness, cracks and potholes, can pose a direct threat to the efficiency, durability and safety of the ERS components this especially in in-pavement inductive or conductive systems. International Roughness Index: The International Roughness Index (IRI) is a quantitative (pavement smoothness) metric used globally which is quoted in metres per kilometre (m/km), and is one of the important quantitative measures of road surface condition [30].

Bangladesh is a good example of the road quality. The 2023–24 Maintenance and Rehabilitation Needs Report provided by the Road Headquarters Department indicates that the state of its National Highway network has significantly improved. These figures represent a sharp change from 2014 to 15, when only 19.46% [31]. This represents a 5 year round of successive efforts to deal with an extensive backlog of maintenance. Although the current situation has been positive, this history highlights the essential nature of maintenance funds to avoid the possibility of going back to poor conditions, which will not benefit any installed ERS [30].

Thailand has a highly developed pavement evaluation technology, like the Hawkeye 2000 system, which combines rutting, macrotexture, and geometry data on the roads to achieve a quality score of 4.40 out of 7 as per the qualitative assessments developed by the World Economic Forum in 2019, meaning that the quality of the surface is high enough to support the electric road system (ERS).

In Indonesia, there are no nationwide data on IRI in detail that can be found easily in the given documentation. However, an analysis of a collector road in Magetan Regency using the Roadroid mobile application reported an average IRI of 4.24/ km that was characterised as moderate to good [32]. The study also revealed the

Table 6.4 Pavement quality comparison across countries

Country	Data source/ indicator	Average IRI (national highways)	% of NH network in "Good" condition	Key pavement distresses
India	General reports	Not available	Not available	Rutting, cracking, potholes (variable)
Bangladesh	RHD (2023)	Not available (categorized)	75.35% (IRI <4.0 m/km)	Surface distress in older sections
Thailand	WEF (2019)	Score: 4.40/7.00 (qualitative)	High (implied)	General wear and tear
Indonesia	Case study (2024)	4.24 m/km (sample road)	Moderate (implied)	Cracks, potholes, deformations

existence of critical areas with moderate to severe damage in the form of cracks and potholes.

Though IRI data on the national highway (NH) network as a whole in India is unavailable, the accelerated growth of the said network often puts more emphasis on the construction of the said network rather than on its maintenance, which contributes to the development of a strong deviation in the quality of roads. Pavement degradation in all four countries is promoted by the tropical climate, with its heavy monsoon rainfall season and high temperatures, as well as the high level of vehicle overloading.

As demonstrated in the analysis, the IRI of a newly built expressway can be perfect at the start; however, the true preparedness is dependent on the institutional and financial ability to maintain it. The latent threat to ERS investment is a significant maintenance backlog, as it has always been the case in Bangladesh. Even a smooth appearance on a surface installed may easily change to a condition where delicate ERS equipment may be at risk, in case the maintenance regime turns out to be inadequate. By extension, the maintenance budgets per year and the ability to decrease the share of the network that is considered to be in poor condition are invaluable but indirect measures of ERS preparedness (Table 6.4).

6.7 ERS Readiness Index (ERI) and Comparative Assessment

In order to incorporate the multidimensional analysis of the infrastructure, policy, and economic determinants into a comparative analysis, this section introduces the Electric Road System Readiness Index (ERI). ERI is a composite measure, which is based on the scale of 0–100, and is designed to provide a quantitative measure of the readiness of India, Bangladesh, Thailand and Indonesia towards the adoption of ERS. The index has five pillars that are equally weighted and represent a critical dimension of readiness.

The ERI framework is structured as follows:

1. **Infrastructure Quality (20%)**: This pillar measures the physical state of the road network. It includes sub-indicators like the density of high-quality, paved highways (NH/Expressways), the ratio of proportion of the network that consists of four lanes or more of the network and the average quality of road-surface in terms of the International Roughness Index (IRI).
2. **Grid Connectivity (20%)**: This pillar assesses the electrical grid's ability to support ERS. Key sub-indicators are the level of transmission and distribution (T&D) losses (scored inversely), a qualitative assessment of grid reliability and stability, and the geographic proximity of high-voltage transmission lines to major freight corridors.
3. **Institutional Readiness (20%)**: This pillar examines how sound the governance and regulatory environment is. It captures the maturity and history of existence of a national public–private partnership (PPP) framework [33]. The perceived efficiency of ancillary processes like land acquisition, and the availability, or unavailability, of ERS-specific technical standards in national engineering codes.
4. **Policy Support (20%)**: This pillar gauges the level of government commitment to transport electrification. Sub-indicators include the strength and comprehensiveness of national electric vehicle (EV) policies, the explicit integration of ERS concepts into national infrastructure master plans (e.g., India's PM Gati Shakti, Thailand's EEC), and stated national carbon reduction targets.
5. **Financial Capacity (20%)**: This pillar measures the economic and financial environment for large-scale infrastructure investment. It includes national infrastructure investment as a percentage of GDP, the average cost of road construction (scored inversely, as higher costs indicate potential inefficiencies), and the country's access to and experience with multilateral development financing.

Based on the data analysed in the preceding sections, scores are assigned and aggregated to produce the final ERI results presented in Table 6.5.

The scores of ERI indicate a clear readiness hierarchy. Thailand [34] stands out as the best-equipped country, as it portrays a balanced scorecard in terms of the quality of infrastructure, stability of the grid and stability of institutions. Its main obstacles

Table 6.5 ERS readiness index (ERI) scores by country

ERI pillar (weight)	India	Bangladesh	Thailand	Indonesia
Infrastructure quality (20%)	12	8	16	11
Grid connectivity (20%)	13	7	15	12
Institutional readiness (20%)	11	9	14	12
Policy support (20%)	17	12	15	14
Financial capacity (20%)	13	8	14	12
Total ERI score (out of 100)	**66**	**44**	**74**	**61**

Source Author analysis (2025)

are associated with the expansion of policy support and monetary investments on ERS alone.

The next in the list is India [35] with a high potential profile but a high execution risk profile. Its performance is also boosted by high levels of top-down policy backing in terms of ambitious national programs and a substantial but fluctuating-quality infrastructure foundation. But it is limited by institutional bottlenecks, especially land acquisition and the efficiency of implementing the projects, and constant gridlocks within certain areas.

The moderate readiness is in Indonesia [31], with the strengths that lie geographically in the island of Java. The nation has an apparent policy drive towards EV production and massive investment in its fundamental toll-road infrastructure. Its preparedness is checked by the archipelagic break-up of its infrastructure and a not-so-well-developed institutional structure to handle complex technology undertakings.

The biggest foundational challenges are in Bangladesh [36]. Its policy intention is increasing, and the recent improvements towards road quality are impressive; however, its underlying flaws in grid stability, the high cost of construction, and insufficient institutional capabilities to support large-scale PPPs lead to a lower overall readiness score.

The comparative advantages and disadvantages of the three countries are captured in the radar chart in Fig. 6.3, which is a clear example of the various development paths that each country needs to follow before it can become a full-fledged ERS-ready country.

6.8 Challenges and Bottlenecks

The route to the implementation of ERS in Emerging Asia is full of serious challenges that touch upon technical, economic, institutional, and social areas. The identification of such bottlenecks requires a sober evaluation of them to come up with effective mitigation measures.

Technical Challenges

The main technical difficulties are associated with the relations between ERS technology and the severe operating conditions.

- **Pavement Integrity**: The high concentration of axle loads (often exceeded illegally) and the extreme climatic conditions (intense monsoon rainfall and high ambient temperatures) contribute to the rate of pavement degradation by increasing unevenness of the surface (high IRI), rutting, and crack propagation, which may damage fragile in-pavement ERS components and reduce the efficiency of power transfer.
- **Grid Instability**: As discussed about Bangladesh, grid instability and T&D losses of significant magnitude are a critical problem [37]. An ERS is a large dynamic

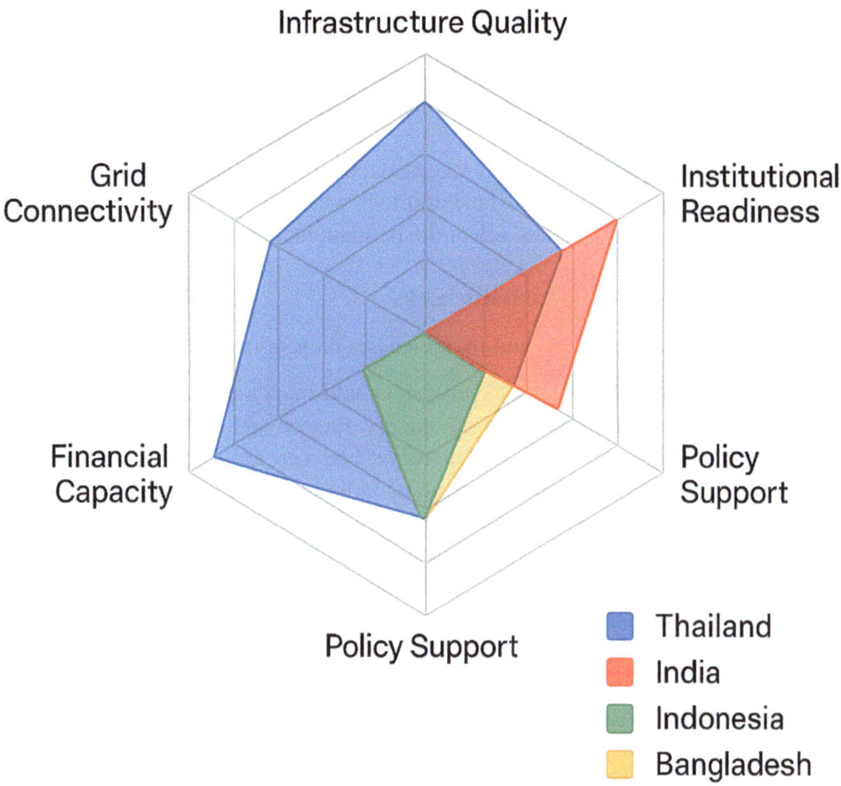

Fig. 6.3 Radar chart—comparative ERS readiness

load; when connected to a weak or unreliable grid, it may cause local voltage drops or frequency variations that affect the system-wide operation. This requires a lot of information on the modernisation of the grid through smart grid technologies in order to guarantee that the grid is capable of carrying such loads.

- **Climate Vulnerability**: Ground-level ERS infrastructure (conductive and inductive) is vulnerable to low-lying and coastal areas, especially in Bangladesh and some areas in Indonesia and Thailand, and climate-resistant designs to high roadbeds or sophisticated waterproofing and drainage are required, further increasing prices.

Economic and Financial Bottlenecks

The economic challenges can be called the most daunting.

- **High Capital Costs**: ERS implementation is a twofold cost: one must upgrade the road system to meet the necessary level and install the costly ERS. The costs of road construction are already very high in the area, and it is reported that Bangladesh possesses one of the highest per-kilometre costs in the world [34].

Funding Limitations

- Public budgets are meagre, and big infrastructural projects often experience budget deficits and overruns [38]. PPPs can be used as a solution, but are not equally effective in all countries.
- **Import Dependence**: At the early stages, the majority of the core ERS technology and components will have to be imported, which will expose the projects to currency fluctuation risks and affect the national balance of payments.

Institutional and Governance Bottlenecks

Institutional inertia and procedural inefficiencies present major obstacles.

- **Fragmented Jurisdictions**: ERS projects should enjoy a smooth flow of coordination between ministries of transport, energy, finance and environment, and state and local governments. Such collaboration between agencies is not a common practice. The PM Gati Shakti platform of India is an explicit effort to solve this bottleneck, but its success remains to be seen [39].
- **Land Acquisition**: The land acquisition process is known to be a painful cause of delay, especially in highly populated nations such as India and Bangladesh [38]. Legal tussles and resistance by the local community can go on to delay projects on a long-term basis.
- **Lack of PPP Frameworks**: In India, PPP frameworks on roads are quite established; however, the frameworks in the other countries are less established and bankable towards investment of this nature, involving a project of this complexity and technological novelty [35].

Social Challenges

The social dimensions of such a large-scale transition must be carefully managed.

- **Informal Freight Sector**: Much of the trucking business in the countries involves small owner-operators and informal businesses. Economic disruption and social inequity may be an unavoidable consequence of the lack of cost-effective solutions to finance the high initial cost of ERS-compatible trucks, thus excluding this group.
- **Land and Livelihood Displacement**: Infrastructure projects are bound to cause the displacement of populations and loss of agricultural land and livelihoods that may result in a substantial backlash in case resettlements and compensation procedures are not addressed fairly and finely [40] (Table 6.6).

6.9 Strategic Interventions and Adoption Pathways

A uniform rollout of Electric Road Systems (ERS) cannot work, as readiness levels are heterogeneous and country-specific conditions vary. Instead, country-specific interventions based on the priorities of national development and stages of adoption

Table 6.6 Major bottlenecks and proposed mitigation strategies

Category	Specific challenge	Proposed mitigation strategy
Technical	Pavement degradation from climate and overloading	Adopt high-strength, climate-resilient pavement materials (e.g., polymer-modified bitumen); enforce axle load regulations strictly
	Grid instability and high T&D losses	Invest in T&D network modernization and smart grid technologies; co-locate ERS with dedicated renewable energy generation (e.g., solar farms)
Economic	Prohibitively high capital expenditure	Phase deployment starting with pilot corridors; leverage multilateral development bank financing; develop innovative PPP models with viability gap funding
Institutional	Land acquisition delays	Utilize integrated planning platforms (e.g., PM Gati Shakti); implement transparent land pooling and compensation models; establish dedicated legal tribunals for disputes
	Lack of inter-ministerial coordination	Create empowered, high-level national ERS task forces with representatives from all relevant ministries to streamline decision-making
Social	Exclusion of informal freight sector	Introduce government-backed financing, leasing, or retrofitting schemes for small truck owners; structure ERS as a service model

are necessary. This will provide easy learning on technology, adaptation of the system, and gradual capacity building with reduced initial financial and technical risks.

Country-Specific Strategic Interventions

- **India**: The best approach to India is to incorporate ERS in its flagship national infrastructure programmes. ERS pilot corridors must be incorporated into the Bharatmala Pariyojana, which aims at streamlining the movement of freight along economic corridors [41]. Ideally, the PM Gati Shakti National Master Plan provides an ideal digital platform to integrate road upgrades, grid connection and development of logistics parks, which is already incorporated in key infrastructure developments.
- **Thailand**: Thailand can use the Eastern Economic Corridor (EEC) as a regulatory sandbox and a zone of focused deployment [42]. The manufacturing and logistics within the high-end segment, as an industrial zone, and closeness to the major deep-sea port, Laem Chabang, are clustered in the EEC. Installation of ERS on major highways in the EEC (e.g., Motorway 7) would be of direct assistance to the national goal of enhancing the logistics capacity and the technological framework of the corridor. Furthermore, passenger applications can be demonstrated with integration with urban Bus Rapid Transit (BRT).
- **Indonesia**: Indonesia has made the Trans-Java Toll Road its strategic focus, and this is the economic backbone of Jakarta, Bandung, and Surabaya. Electrification of this line would have the strongest effect on the national economy and emission profile. This intervention must be very much in line with the industrial policy

of Indonesia so that it is possible to build a regional centre of electric vehicle (EV) and battery manufacturing, taking advantage of the fact that Indonesia is one of the largest producers of nickel, which can be used to develop a vertically integrated ecosystem of raw materials to electrified logics [43].

- **Bangladesh**: In the case of Bangladesh, the plan should be an action plan that has a basis in upgrading infrastructure. The first initiative must be focused solely on the Dhaka-Chattogram line, which is the most crucial line of national trade. A project on this route by ERS should be a part of a larger package that should involve specific investment in the modernisation of the electrical grid in the corridor to guarantee the reliability currently absent.40 A second phase may be proposed on the axis connecting the Benapole land port.

Phased Adoption Pathway

A long-term, two-phase approach is recommended for the region:

- **Phase 1 (2025–2035): Pilot Corridors and Institutional Framework Development**. The focus of this decade should be on the implementation of small-scale pilot projects (50,100 km) on the above strategic corridors. The main goals of this step are to examine the technology in the conditions of the local climate and traffic, gather data on its work, and optimise the economic model. At the same time, governments need to focus on the development of the so-called soft infrastructure: national technical standards of ERS, adaptation of the models of public–private partnership (PPP) and concession agreements, the creation of domestic technical and regulatory capacity, and the development of mechanisms to fund vehicle conversion or purchase.
- **Phase 2 (2035–2045): Scaled Expansion and Network Integration**. Using the experience of successful pilots, this step would involve the gradual extension of ERS to extend along the entire primary economic corridors. This would drive towards the interoperability of different ERS segments and, possibly, across borders to enable a domestic supply chain of ERS components to save on costs and reliance on imports, and a profound interconnection of the ERS network with large-scale renewable energy sources to make the logistics chain truly green.

Figure 6.4 presents a conceptual roadmap illustrating this phased approach with country-specific milestones.

A timeline-based infographic from 2025 to 2045. It features four parallel horizontal lanes, one for each country. Key milestones are placed along each timeline. For India: "2028: Delhi-Mumbai ERS Pilot (100 km) launched," "2032: National ERS Standards Published," "2040: Golden Quadrilateral ERS Expansion." For Thailand: "2027: EEC Pilot (Laem Chabang-Bangkok) Operational," "2035: Expansion to Northern Economic Corridor." For Indonesia: "2029: Trans-Java ERS Pilot (Jakarta-Bandung)," "2033: National Battery & ERS Integration Policy," "2042: Trans-Java Full Electrification." For Bangladesh: "2030: Dhaka-Chattogram ERS Pilot + Grid Upgrade," "2040: Expansion to Benapole Corridor."

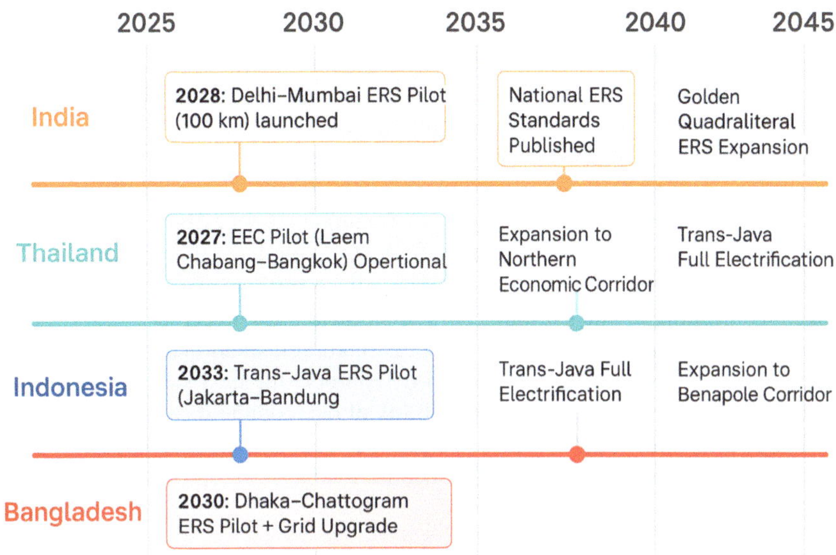

Fig. 6.4 Strategic roadmap for ERS deployment in emerging Asia

6.10 Conclusion

The chapter has also presented a comparative evaluation of the infrastructural, institutional, and policy preparedness of Electric Road Systems in India, Bangladesh, Thailand, and Indonesia in a comprehensive manner. The analysis is made by the ERS Readiness Index (ERI), which clearly, but in a complicated manner, shows the picture synthesised. Thailand is the best-prepared country in a holistic sense because its infrastructure is of high quality, and the grid is very strong. India has huge potential due to the good policy ambition and scale, but is still hindered by major institutional barriers. The preparedness of Indonesia is geographically focused and strictly connected with its industrial policy in the world, and Bangladesh faces the most basic issues, which are grid reliability and high infrastructure development expenses.

The key conclusion of this evaluation is that the key obstacles to the adoption of ERS in emerging Asia are not grounded in road engineering but in institutional deficiencies and gridlocks. Efforts to build roads will be less significant to the success of ERS, but rather the capability to have a well-coordinated effort amongst the ministers, the ability to purchase the land in a manner that is bankable, the ability to establish bankable PPP structures and most importantly, the capacity to modernise the electrical transmission and distribution networks. The reliability of the grid and not the capacity of the generation is the real knuckle bone of ERS viability.

The solution to these preparedness gaps requires a strategic step-wise intervention depending on the circumstances of the country. Policymakers can establish strong synergies and drive rapid development by including ERS planning in the current national development programmes such as the Bharatmala and PM Gati in India, the Eastern Economic Corridor in Thailand, and the Trans Java in Indonesia. The pilot corridors will play a vital role in the beginning, as they will be important in technology validation, risk mitigation, and capacity building.

In addition, there is also a strong argument for regional cooperation. Institutions like the Asian Development Bank or UNESCAP could facilitate a platform through which technical standards, pilot project information and policy best practices could be shared so that different countries could learn about each other and align their practices. This type of cooperation may assist in the de-risking of a private investment as well as speeding up the learning curve of all the parties involved.

With this elaborate background of infrastructural/institutional preparedness in place and the key obstacles having been determined, the next step in the analysis process is to examine the economic feasibility of overcoming these challenges. The next chapter will thus come up with a stringent cost–benefit examination and examine novel funding frameworks that would be needed to finance the shift to electrified road freight in Asia.

References

1. Wikipedia. Roads in India. https://en.wikipedia.org/wiki/Roads_in_India. Accessed 16 Oct 2025
2. Wikipedia. National highways of India. https://en.wikipedia.org/wiki/National_highways_of_India. Accessed 16 Oct 2025
3. Asian Development Bank (ADB). Infrastructure. https://www.adb.org/sites/default/files/publication/27728/infrastructure.pdf. Accessed 16 Oct 2025
4. Wikipedia. Transport in Thailand. https://en.wikipedia.org/wiki/Transport_in_Thailand. Accessed 16 Oct 2025
5. Wikipedia. Transport in Indonesia. https://en.wikipedia.org/wiki/Transport_in_Indonesia. Accessed 16 Oct 2025
6. Wikipedia. List of roads in Bangladesh. https://en.wikipedia.org/wiki/List_of_roads_in_Bangladesh. Accessed 16 Oct 2025
7. xMap AI. Thailand. Road traffic data. https://www.xmap.ai/data-catalogs/thailand-road-traffic-data. Accessed 16 Oct 2025
8. Statistics Indonesia (BPS). Length of road by level of government responsibility. https://www.bps.go.id/en/statistics-table/2/NTAjMg==/length-of-road-by-level-of-government-responsibility.html. Accessed 16 Oct 2025
9. Press Information Bureau (PIB). Year end review 2023—Ministry of Road Transport and Highways. https://www.pib.gov.in/PressReleaseIframePage.aspx?PRID=1993425. Accessed 16 Oct 2025
10. Roads and Highways Department (RHD). Online road network. https://www.rhd.gov.bd/OnlineRoadNetwork/Default.asp. Accessed 16 Oct 2025
11. Wikipedia. Thai highway network. https://en.wikipedia.org/wiki/Thai_highway_network. Accessed 16 Oct 2025
12. Wikipedia. Asian highway network. https://en.wikipedia.org/wiki/Asian_Highway_Network. Accessed 16 Oct 2025
13. Wikipedia. Transport in Bangladesh. https://en.wikipedia.org/wiki/Transport_in_Bangladesh. Accessed 16 Oct 2025

14. United Nations ESCAP. Asian highway database—ESCAP repository. https://repository.unescap.org/items/eb868e1e-6656-4eba-8552-336302bc24a0. Accessed 16 Oct 2025
15. Scribd. Indian road standards and guidelines. https://www.scribd.com/document/381992796/Indian-Standards-List-Roads. Accessed 16 Oct 2025
16. Ministry of Foreign Affairs, Thailand. General information of international road infrastructure development for ASEAN and BIMSTEC networks (phase 2). https://image.mfa.go.th/mfa/0/GH2PYnujXi/%E0%B9%80%E0%B8%AD%E0%B8%81%E0%B8%AA%E0%B8%B2%E0%B8%A3/International_Road_Infrastructure_Development.pdf. Accessed 16 Oct 2025
17. Directorate General of Highways (Bina Marga), Indonesia. Standar Pedoman. https://binamarga.pu.go.id/index.php/konten/en/standar-pedoman/. Accessed 16 Oct 2025
18. ResearchGate. Siapkah Bidang Bina Marga Menerapkan Pembangunan Jalan yang Berkelanjutan. https://www.researchgate.net/publication/366465091_Siapkah_bidang_Bina_Marga_menerapkan_Pembangunan_Jalan_yang_Berkelanjutan. Accessed 16 Oct 2025
19. Press Information Bureau (PIB). *Year End Review 2024 – Ministry of Road Transport and Highways.* Available at: https://www.pib.gov.in/PressReleasePage.aspx?PRID=2089243. Accessed 16 Oct 2025.
20. World Bank. *World Development Indicators | DataBank.* Available at: https://databank.worldbank.org/id/a799c980. Accessed 16 Oct 2025.
21. World Bank. *Electric Power Transmission and Distribution Losses (kWh) – Glossary | DataBank.* Available at: https://databank.worldbank.org/metadataglossary/world-development-indicators/series/EG.ELC.LOSS.KH. Accessed 16 Oct 2025.
22. The Financial Express. *Power Generation Capacity Crosses 30,000 MW in 2023.* Available at: https://today.thefinancialexpress.com.bd/print/power-generation-capacity-crosses-30000mw-in-2023-1704303617. Accessed 16 Oct 2025.
23. PLN Indonesia. *Statistics PLN 2023.* Available at: https://web.pln.co.id/statics/uploads/2024/08/Bk-Statistik-PLN-2023-English-31.7.24.pdf. Accessed 16 Oct 2025.
24. Press Information Bureau (PIB). *Intermodal Connectivity for Freight Traffic.* Available at: https://www.pib.gov.in/PressReleaseIframePage.aspx?PRID=2116158. Accessed 16 Oct 2025.
25. DST News. *Dedicated Freight Corridors: Transforming Rail Logistics with Record Growth.* Available at: https://dst.news/news/dedicated-freight-corridors-transforming-rail-logistics-with-record-growth/. Accessed 16 Oct 2025.
26. CEIC Data. *Thailand Road Freight Transport – Economic Indicators.* Available at: https://www.ceicdata.com/en/thailand/road-and-rail-freight-transport-annual/road-freight-transport. Accessed 16 Oct 2025.
27. ResearchGate. *Investigating the Transportation Economics of the Eastern Economic Corridor in Thailand.* Available at: https://www.researchgate.net/publication/396277546_Investigating_the_Transportation_Economics_of_the_Eastern_Economic_Corridor_in_Thailand. Accessed 16 Oct 2025.
28. Mordor Intelligence. *Indonesia Road Freight Transport Market Size & Share Analysis.* Available at: https://www.mordorintelligence.com/industry-reports/indonesia-road-freight-transport-market. Accessed 16 Oct 2025.
29. ARRB Systems. *Thailand's Highways: A Case Study in Advanced Road Maintenance.* Available at: https://arrbsystems.com/case-studies/department-of-highways/. Accessed 16 Oct 2025.
30. Journal of Social Research. *Analysis of Pavement Condition Using the International Roughness Index (IRI) Method with the Roadroid Application on the Genengan–Lembeyan Road.* Available at: https://ijsr.internationaljournallabs.com/index.php/ijsr/article/download/2469/1357/12554. Accessed 16 Oct 2025.
31. Wikipedia. *Public–Private Partnerships in India.* Available at: https://en.wikipedia.org/wiki/Public%E2%80%93private_partnerships_in_India. Accessed 16 Oct 2025.
32. Bureau of Energy Efficiency (BEE). *National Electric Mobility Mission Plan.* Available at: https://evyatra.beeindia.gov.in/central-govt-initiative-details/dhi-2/. Accessed 16 Oct 2025.
33. BizBD Review. *Bangladesh Charts Roadmap for EV Revolution with New Policy.* Available at: https://bizbdreview.com/bangladesh-charts-roadmap-for-ev-revolution-with-new-policy/. Accessed 16 Oct 2025.

34. Eco-Business. *ADB and World Bank Launch Financing Initiative to Boost ASEAN Power Grid.* Available at: https://www.eco-business.com/news/adb-world-bank-launch-financing-initiative-to-boost-asean-power-grid/. Accessed 16 Oct 2025.
35. Wikipedia. *Plug-in Electric Vehicles in Thailand.* Available at: https://en.wikipedia.org/wiki/Plug-in_electric_vehicles_in_Thailand. Accessed 16 Oct 2025.
36. Wikipedia. *Bharatmala.* Available at: https://en.wikipedia.org/wiki/Bharatmala. Accessed 16 Oct 2025.
37. World Bank. *Bangladesh – Rural Electricity Transmission and Distribution Project: Reliable Power Supply for Rural Bangladesh.* Available at: https://documents.worldbank.org/en/publication/documents-reports/documentdetail/322001568287322576/bangladesh-rural-electricity-transmission-and-distribution-project-reliable-power-supply-for-rural-bangladesh. Accessed 16 Oct 2025.
38. Wikipedia. *Pradhan Mantri Gati Shakti.* Available at: https://en.wikipedia.org/wiki/Pradhan_Mantri_Gati_Shakti. Accessed 16 Oct 2025.
39. Antara News. *Minister Stresses Commitment to Enhancing Sustainable Infrastructure Governance.* Available at: https://en.antaranews.com/news/385957/minister-stresses-commitment-to-enhancing-sustainable-infra-governance. Accessed 16 Oct 2025.
40. MagicBricks. *Bharatmala Project: Routes, Map, Details, and Latest News.* Available at: https://www.magicbricks.com/blog/bharatmala-project-map-details-news/128887.html. Accessed 16 Oct 2025.
41. Railway Supply. *India's Freight Corridors: Transforming Rail Logistics with Record Growth.* Available at: https://www.railway.supply/indias-freight-corridors-transforming-rail-logistics-with-record-growth/. Accessed 16 Oct 2025.
42. Wikipedia. *International Roughness Index.* Available at: https://en.wikipedia.org/wiki/International_roughness_index. Accessed 16 Oct 2025.
43. Bangkok Post. *Exat Eyes Four New Road Projects.* Available at: https://www.bangkokpost.com/thailand/general/2745011/exat-eyes-4-new-road-projects. Accessed 16 Oct 2025.
44. Asian Transport Observatory. https://asiantransportobservatory.org/. Accessed 16 Oct 2025
45. World Bank. Private participation in infrastructure (PPI)—2020 annual report. https://ppi.worldbank.org/content/dam/PPI/documents/PPI_2020_AnnualReport.pdf. Accessed 16 Oct 2025
46. Asian Development Bank (ADB). Asian transport outlook database. ADB Data Library. https://data.adb.org/dataset/asian-transport-outlook-database. Accessed 16 Oct 2025
47. World Bank. Roads total and paved per country—world development indicators. DataBank. https://databank.worldbank.org/id/124b6f90. Accessed 16 Oct 2025
48. Wikipedia. Department of rural roads. https://en.wikipedia.org/wiki/Department_of_Rural_Roads. Accessed 16 Oct 2025
49. Roads and Highways Department, Bangladesh. Regional road connectivity—Bangladesh perspective. https://rthd.portal.gov.bd/sites/default/files/files/rthd.portal.gov.bd/page/9b7d5069_9a30_4fe4_85ec_9db4f2f3966b/rrcbp.pdf. Accessed 16 Oct 2025
50. RoadVision.ai. Understanding IRC code: key guidelines for road construction and maintenance in India. https://www.roadvision.ai/blog/understanding-irc-code-key-guidelines-for-road-construction-and-maintenance-in-india. Accessed 16 Oct 2025
51. RoadVision.ai. IRC Code SP:15-1996—guidelines for highway construction and maintenance. https://www.roadvision.ai/blog/irc-code-sp-15-1996---guidelines-for-highway-construction-and-maintenance. Accessed 16 Oct 2025
52. Devex. Department of highways (DOH)—Thailand. https://www.devex.com/organizations/department-of-highways-doh-thailand-159358. Accessed 16 Oct 2025
53. Department of Highways, Thailand. Information memorandum: public private partnerships for operation and maintenance (O&M) Bang Pa-In—Nakhon Ratchasima (M6). http://www.doh-motorway.com/wp-content/uploads/pdf/Final-Information-Memorandum-M6-31012018.pdf. Accessed 16 Oct 2025
54. Directorate General of Highways (Bina Marga), Indonesia. Road development: pay attention to environmental aspects. https://binamarga.pu.go.id/index.php/berita/road-development-pay-attention-to-environmental-aspects. Accessed 16 Oct 2025

55. ResearchGate. Pavement design comparative analysis using Bina Marga design method and pavement design manual (case study: Trisakti Port—Liang Anggang). https://www.researchgate.net/publication/327501896_Pavement_Design_Comparative_Analysis_Using_Bina_Marga_Design_Method_Pt_T-01-2002-B_And_Pavement_Design_Manual_No04SeDb2017_Case_Study_Trisakti_Port_-_Liang_Anggang. Accessed 16 Oct 2025
56. Roads and Highways Department, Bangladesh. Standard tender documents—Section 6.3 (June 2016). https://rhd.portal.gov.bd/sites/default/files/files/rhd.portal.gov.bd/page/2022924c_9c9a_4c3d_bd9b_205e36f4c75e/Standard_Tender_%20Documents_Section_3_June_2016_1.pdf. Accessed 16 Oct 2025
57. 3C Training Bangladesh. *Navigating the Road to Progress: Road Design and Construction in Bangladesh.* Available at: https://3ctrainingbd.com/road-design-construction-in-bangladesh/. Accessed 16 Oct 2025.
58. World Bank. *Electric Power Transmission and Distribution Losses (% of Output) – United States.* Available at: https://data.worldbank.org/indicator/EG.ELC.LOSS.ZS?locations=US. Accessed 16 Oct 2025.
59. Fibre2Fashion. *Bangladesh's Power Capacity Soars by 5,000 MW in 2023.* Available at: https://www.fibre2fashion.com/news/textile-news/bangladesh-s-power-capacity-soars-by-5-000-mw-in-2023-292195-newsdetails.htm. Accessed 16 Oct 2025.
60. Institute for Energy Economics and Financial Analysis (IEEFA). *How to Make Bangladesh's Power Sector Sustainable.* Available at: https://ieefa.org/resources/how-make-bangladeshs-power-sector-sustainable. Accessed 16 Oct 2025.
61. World Business Outlook. *A Closer Look at the Power Sector of the Land of White Elephants.* Available at: https://worldbusinessoutlook.com/a-closer-look-at-the-power-sector-of-the-land-of-white-elephants/. Accessed 16 Oct 2025.
62. Scribd. *Dedicated Freight Corridor – East West, North South and East Coast Corridors.* Available at: https://www.scribd.com/document/825434385/dedicated-freight-corridor-East-west-corridor-East-west-dedicated-freight-corridor-north-south-corridor-east-coast-corridor. Accessed 16 Oct 2025.
63. xMap AI. *India | Road Traffic Data.* Available at: https://www.xmap.ai/data-catalogs/india-road-traffic-data. Accessed 16 Oct 2025.
64. Wikipedia. *Dedicated Freight Corridors in India.* Available at: https://en.wikipedia.org/wiki/Dedicated_freight_corridors_in_India. Accessed 16 Oct 2025.
65. Mordor Intelligence. *Logistics Industry in Bangladesh – Market Companies & Statistics.* Available at: https://www.mordorintelligence.com/industry-reports/bangladesh-freight-and-logistics-market. Accessed 16 Oct 2025.
66. Royal Thai Embassy, Doha. *Thailand's Eastern Economic Corridor (EEC) – Factsheet.* Available at: https://doha.thaiembassy.org/en/content/eec-factsheet?cate=6033aa0b2ab758786f4b3912. Accessed 16 Oct 2025.
67. Wikipedia. *Eastern Economic Corridor.* Available at: https://en.wikipedia.org/wiki/Eastern_Economic_Corridor. Accessed 16 Oct 2025.
68. Roads and Highways Department, Bangladesh. *Maintenance and Rehabilitation Needs Report 2023.* Available at: https://rhd.portal.gov.bd/sites/default/files/files/rhd.portal.gov.bd/page/8dde7a5d_251e_4ba4_bcd0_82746bb33478/2023-06-06-06-12-d233e504c7732c62d00055b74fdb3a9b.pdf. Accessed 16 Oct 2025.
69. TheGlobalEconomy.com. *Thailand Roads Quality – Data and Chart.* Available at: https://www.theglobaleconomy.com/Thailand/roads_quality/. Accessed 16 Oct 2025.
70. King & Wood Mallesons (KWM). *Public–Private Partnerships in Bangladesh.* Available at: https://www.kwm.com/au/en/insights/latest-thinking/public-private-partnerships-in-bangladesh.html. Accessed 16 Oct 2025.
71. PwC Indonesia. *Infrastructure Development: Promising Opportunities for PPP Projects.* Available at: https://www.pwc.com/id/en/media-centre/infrastructure-news/april-2025/infrastructure-development-promising-opportunities-for-ppp-projects.html. Accessed 16 Oct 2025.
72. International Council on Clean Transportation (ICCT). *Policies to Help Indonesia's New President Champion Electric Vehicles.* Available at: https://theicct.org/policies-to-help-indonesias-new-president-champion-evs-aug25/. Accessed 16 Oct 2025.

73. Press Information Bureau (PIB). *Press Note Details.* Available at: https://www.pib.gov.in/PressNoteDetails.aspx?id=155094&NoteId=155094&ModuleId=3. Accessed 16 Oct 2025.
74. Parliament of India. *Sansad Paper – Cost Estimates of Road Construction.* Available at: https://sansad.in/getFile/annex/218/Au3050.pdf?source=pqars#:~:text=14%20crore%20per%20km%20for,acquisition%20and%20other%20preconstruction%20expenses. Accessed 16 Oct 2025.
75. Asia News Network. *The Price of Expensive Development.* Available at: https://asianews.network/the-price-of-expensive-development/. Accessed 16 Oct 2025.
76. Global Highways. *Major Indonesian Highway Project on Track.* Available at: https://www.globalhighways.com/wh10/news/major-indonesian-highway-project-track. Accessed 16 Oct 2025.
77. SKIL Global. *Five Common Problems in Indian Road Construction and How to Fix Them.* Available at: https://www.skilglobal.com/5-common-problems-in-indian-road-construction-and-how-to-fix-them/. Accessed 16 Oct 2025.
78. Daily Sun. *World Bank Finds Road Construction Cost in Bangladesh World's Highest.* Available at: https://www.daily-sun.com/post/235348/WB-finds-road-construction-cost-in-Bangladesh-world%E2%80%99s-highest. Accessed 16 Oct 2025.
79. Prothom Alo English. *Highway Construction Costs Higher in Bangladesh than Neighbours.* Available at: https://en.prothomalo.com/bangladesh/x820tijuk0. Accessed 16 Oct 2025.
80. Government of India. *PM Gati Shakti – National Master Plan for Multi-modal Connectivity.* Available at: https://www.india.gov.in/spotlight/pm-gati-shakti-national-master-plan-multi-modal-connectivity. Accessed 16 Oct 2025.
81. The Economic Times. *Highway Projects Worth ₹4 Lakh Crore Face Delays; 200 More Worth ₹1.6 Lakh Crore Await Construction Start.* Available at: https://m.economictimes.com/news/economy/infrastructure/highway-projects-worth-4-lakh-crore-face-delays-200-more-worth-1-6-lakh-crore-await-construction-start/articleshow/123139930.cms. Accessed 16 Oct 2025.
82. Public–Private Infrastructure Advisory Facility (PPIAF). *Public–Private Partnership (PPP) in Thailand.* Available at: https://www.ppiaf.org/documents/4225. Accessed 16 Oct 2025.
83. Internet Geography. *What Are the Positive and Negative Impacts of Rapid Development in India?* Available at: https://www.internetgeography.net/topics/what-are-the-positive-and-negative-impacts-of-rapid-development-in-india/. Accessed 16 Oct 2025.
84. ResearchGate. *Bangladesh's Structural Transformation: The Role of Infrastructure.* Available at: https://www.researchgate.net/publication/321276298_Bangladesh's_Structural_Transformation_The_Role_of_Infrastructure. Accessed 16 Oct 2025.
85. PIARC. *Socio-Economic Impact of Transport Infrastructure: A Bangladesh Case Study (Abu Saleh Md. Nuruzzaman).* Available at: https://proceedings-paris2007.piarc.org/ressources/files/5/IP514-Nuruzzaman-E.pdf. Accessed 16 Oct 2025.
86. United Nations Development Programme (UNDP). *Infrastructure Development and Women's Rights in Indonesia.* Available at: https://files.acquia.undp.org/public/migration/id/a73e36abffa45ef5d593609c39c224c1234b598073f3d1b21ae26b471c9df361.pdf. Accessed 16 Oct 2025.
87. NimbusPost. *Bharatmala Project: Meaning, Roadmap and Impact on Indian Logistics.* Available at: https://nimbuspost.com/blog/bharatmala-project-meaning-roadmap-impact-on-indian-logistics. Accessed 16 Oct 2025.
88. Press Trust of India (PTI). *Government Opens PM Gati Shakti Portal to Private Sector.* Available at: https://www.ptinews.com/story/business/govt-opens-pm-gati-shakti-portal-to-pvt-sector/3003215. Accessed 16 Oct 2025.
89. International Institute for Sustainable Development (IISD). *Indonesian Electric Vehicle Boom: A Temporary Trend or a Long-Term Vision?* Available at: https://www.iisd.org/articles/deep-dive/indonesian-electric-vehicle-boom-temporary-trend-or-long-term-vision. Accessed 16 Oct 2025.

Chapter 7
Cost–Benefit Analysis and Financing Models for Electric Road Systems (ERS)

7.1 Introduction: The Economic Rationale for Dynamic Charging

Electric Road Systems (ERS) are an inherent paradigmatic transformation of the process of vehicle energy delivery, which ceases discrete refuelling or recharging events, moving to in-motion power transfer. This chapter examines the complex economic context of ERS, thereby defining the main tension that shapes its business case: the substantial capital outlay (CapEx) to build shared public infrastructure and the possibility of significant, system-wide cost savings and overall benefits to society in the long run [1].

The major driver behind the assessment of such a transformative infrastructure project is the urgent need to decarbonise the heavy-duty vehicle (HDV) industry. Although HDVs contribute only a small fraction of the total vehicle fleet, they emit a disproportionate share of transport-related greenhouse gases and making their electrification the necessary step to achieve national and international climate goals [2]. ERS, also known informally as dynamic charging, provides a feasible solution to the issues associated with electrifying long-haul freight, which cannot be easily solved with stationary charging. The key question to be addressed by this analysis is as follows: under what conditions can ERS be the most economical and scalable technology in the case of HDV decarbonisation compared to other technologies?

This chapter is structured in such a way that it is an evaluation of the economic viability of ERS. It begins with a microeconomic examination of the entire life-cycle expenses related to the ERS infrastructure and vehicles, which use it. It then goes on to consider a macroeconomic analysis of the welfare of the society through the formal Cost–Benefit Analysis (CBA), including externalities that include environmental and civilian health effects. Lastly, the chapter discusses the practical issues of the project delivery process, covering a variety of models of financing and architectures of investments necessary to finance and implement ERS networks on a large scale. This overall strategy will provide a clear guide to the policymakers, investors,

R. Agnihotri, *Electrified Road Systems (ERS) for Emerging Asia*, Springer Tracts on Transportation and Traffic 24, https://doi.org/10.1007/978-3-032-16927-3_7

and infrastructure planners who may consider investing in this next generation of transport technology.

7.1.1 A Life-Cycle Cost Framework for ERS

An intense evaluation of the ERS feasibility begins with the bottom-up evaluation of all the expenditures during the life of the project. Using well-known frameworks, like the Levelized Cost per X-mile (LCXM), which is a metric used to measure the cost of a life-cycle of an item of transportation service provided (e.g., per ton-mile), ensures that both capital and operational costs are analysed simultaneously.

7.1.2 Deconstructing Capital Expenditures (CapEx)

Large start-up capital is one of the main characteristics of ERS projects. This financial structure is typical of most green infrastructure assets, which actually transform the economic risk profile of volatile and long-term operational costs, like fluctuations in fossil fuel prices, into predictable and initial investment costs.

The major infrastructure investment items include large civil works to modify the roads, install the power transfer systems (including overhead catenary lines, in-road conductive rails, and inductive coils), and the installation of the necessary power electronics, grid-connection substations, and advanced control and monitoring systems. The cost per kilometre of this infrastructure will change significantly based on the technology adopted, the local geographical aspects and the level of grid reinforcement to meet the high-power demand of the system. Published estimates show this range, with large-scale projects ranging between € 0.4–1.1 million per kilometre to put up a Conductive Power Transfer (CPT) system and € 0.09–1.35 million per kilometre to put up a Wireless Power Transfer (WPT) system, equivalent to about € 0.9 million–1.33 million per kilometre, respectively. In addition, a full CapEx analysis should also take into consideration the frequently neglected external costs, such as purchasing land to locate substations, and environmental mitigation costs incurred.

7.1.3 Analysing Operational Expenditures (OpEx)

Although CapEx in ERS is high, the potential benefit of ERS is that there are low but reasonably constant operating expenses as opposed to the traditional fossil-based transport systems. The procurement of electricity is the greatest OpEx element. The prices of electricity tend to be less volatile than those of diesel, making the adoption of such technology less susceptible to market changes, which will have a direct impact on the overall quantity of energy that will be required by the fleet operator to buy [3]. The effectiveness of the technology adopted in the ERS, be it conductive or inductive, will directly affect the amount of energy that will be bought [4].

Another important OpEX element is maintenance. The infrastructure of the ERS itself needs periodic checks and maintenance of power lines or in-road rails, the routine maintenance of the substations, and the constant updating of the software of control systems. One major difference arises between technologies: conductive systems, which rely on physical contact, are prone to a higher degree of mechanical wear and will probably incur higher costs of maintenance than non-contact inductive systems [5], but these infrastructure maintenance costs have to be counterbalanced by the large maintenance savings that are achieved at the vehicle level. Electric trucks have significantly fewer moving components than diesel ones, which therefore translates into significantly lower services, maintenance, and repair costs over the life of the vehicle, a significant factor in the Total Cost of Ownership (TCO) equation [6]. Lastly, administrative and billing costs such as staff to monitor the system, support customers, and manage transactions and payment platforms should also be included in total OpEx [7].

7.2 Comparative Cost Analysis of ERS Technologies

The choice between conductive and inductive ERS is a fundamental strategic decision with significant long-term implications for cost, performance, and market applicability. Conductive technologies, including both overhead catenary lines and in-road rails, are generally considered the most economically attractive option at present. This is due to their higher power transfer efficiency and greater technological maturity, stemming from decades of use in the rail sector. Studies estimate that the driving cost for a heavy-duty truck using a Conductive Power Transfer (CPT) system ranges from $0.21 to $0.67 per kilometre, with the cost decreasing as traffic volume and system utilization increase. However, overhead systems are functionally limited to taller vehicles such as trucks and buses, which constrains their potential user base.

Inductive technology, or in-motion Wireless Power Transfer (WPT), is generally expected to have higher installation costs than conductive solutions. This is reflected in higher estimated driving costs, which range from $0.22 to $1.03 per kilometer. The higher cost is partly a function of lower energy efficiency, as some power is inevitably lost during the wireless transfer process. Despite these drawbacks, inductive ERS offers compelling advantages. The absence of exposed live parts provides an inherent safety benefit, while the lack of physical contact reduces mechanical wear and associated maintenance. Crucially, because the technology is embedded in the road surface, it can be used by a wide range of vehicles, including passenger cars, which could significantly increase system utilization and improve the overall business case over the long term.

This technological decision is therefore path-dependent and dictates the entire future market structure. An investment in overhead catenary systems is an optimized solution for the immediate and pressing problem of decarbonizing the HDV sector,

Table 7.1 Comparative life-cycle cost analysis of ERS technologies (per dual-lane km)

Technology type	Estimated CapEx (€/km)	Estimated annual OpEx (€/km/year)	Power transfer efficiency (%)	Technology readiness level (TRL)	Key advantages	Key disadvantages
Overhead conductive	€1.0 M–€2.5 M	Low–Medium	>90%	TRL 6–8	High efficiency; mature technology; high power transfer capability	Visually intrusive; limited to tall vehicles (trucks, buses); mechanical wear
In-road conductive	€1.5 M–€3.0 M	Medium	>90%	TRL 4–7	High efficiency; can support cars and trucks	Exposed to elements; safety concerns with live parts; mechanical wear
Inductive WPT	€2.0 M–€4.0 M	Low	80–90%	TRL 3–6	High safety (no exposed parts); low mechanical wear; supports all vehicle types	Higher cost; lower efficiency; lower power transfer capability; less mature

Note Cost estimates are indicative and vary based on project scope, geography, and grid requirements. TRL reflects the maturity of dynamic, in-motion systems for public roads
Data synthesized from [1, 4, 5, 8–12]

leveraging a mature and efficient technology. In contrast, an investment in in-road inductive systems, while more expensive and less efficient today, represents a bet on a more universal charging backbone capable of serving a broader, multi-modal vehicle fleet in the future. This choice has multi-decade consequences, forcing policymakers to decide whether they are solving for the *HDV problem* specifically or creating a foundational infrastructure for a fully electrified road transport ecosystem (Table 7.1).

7.3 Socio-Economic Cost–Benefit Analysis (CBA)

While a life-cycle cost analysis is concerned with the direct financial inputs and outputs of a project, a socio-economic Cost–Benefit Analysis (CBA) takes the analysis to a wider societal level. A more fundamental question it attempts to answer is "Is this investment a net gain to the welfare of society at large?" This requires the quantification not only of direct economic effects, but also of an extensive range of external costs and benefits not normally included in a private financial model.

7.3.1 Methodological Foundations

A CBA for a large-scale public infrastructure project like the Electric-Road-System (ERS) will have to be performed from a societal perspective, which is fundamentally different to a private business case. The analysis is based on proven methodologies for transport schemes, using such metrics as Net Present Value (NPV) (the total discounted value of all future benefits less the total discounted value of all future costs); and the Benefit–Cost Ratio (BCR) to establish overall viability [13]. In the framework, a crucial difference between the two is that taxes and subsidies are accounted for differently. In the society CBA, these are treated as transfers between different actors within the economy (e.g. from taxpayers to project developers/users) as opposed to net costs or benefits to society, but included to reflect distributional consequences [13]. The whole analysis is based on the social discount rate, which is the value society places on the future benefits compared with the present costs.

7.3.2 Direct and Indirect Economic Benefits

The greatest economic advantage of ERS is to optimise the entire electric automobile ecosystem. By delivering uninterrupted power to key corridors, ERS essentially solves the basic conflict that exists between the range of the electric vehicle and its price. This allows the use of vehicles with much smaller, cheaper and lighter on-board batteries, which is currently the single most expensive part of an EV [1]. Battery downsizing creates a chain of positive economic impacts, including reducing the upfront cost to buy a new vehicle, reducing vehicle weight, which boosts energy efficiency, and decreasing the demand on the economy-wide supply of critical minerals such as lithium and cobalt, which in turn reduces the risk of supply chain disruptions [14]. Several analysts believe net savings created by this battery downsizing can outweigh the entire cost of the ERS infrastructure and make this system economically attractive from this perspective alone [15].

Apart from the vehicle, ERS produces important operational efficiencies for the logistics industry. These include direct savings from reduced and more stable energy costs (electricity vs. diesel) [16], reduced vehicle maintenance and repair costs due to more simplified vehicle mechanics of EV [17], as well as elimination of unproductive driver and vehicle downtime that would otherwise be spent at stationary charging stations [3].

On a macroeconomic level, the use of ERS networks can have a positive economic stimulus effect. The construction, manufacture and operation of such systems lead to the creation of skilled jobs in areas such as civil and electrical engineering, project management and high-technology manufacturing [18]. In addition, by changing the source of energy for freight transport from imported oil to domestically generated electricity, ERS substantially increases the national energy security and insulates the logistics sector from the volatility of global oil markets [13]. By doing this, a variable cost of operation is converted into a fixed cost that can be amortised, which in turn creates more macroeconomic stability (Table 7.2).

7.4 Monetising Externalities

A good CBA should also include externalities—costs and benefits to society which are not captured in market prices. A major advantage of ERS is the elimination of the negative externalities of diesel-powered freight. By enabling the switch to electric vans, ERS will lead to a significant reduction of greenhouse gas emissions (CO2e) as well as local air pollutants such as nitrogen oxides (NOx) and particulate matter (PM2.5). Appropriate methodologies, for instance, those offered by the U.S. Environmental Protection Agency (EPA) or the GHG Protocol, can be applied to assign a monetary value to these avoided emissions, which is a real benefit to society [25].

The emission reductions are directly converted into important public health benefits. Reductions in the size of the air pollutant load in high traffic corridors lead to reductions in the rates of respiratory ailments such as asthma and bronchitis, especially among vulnerable populations. Indeed, the societal benefit case for ERS includes a significant economic value component by avoiding healthcare costs and reducing lost productivity due to improved health [18]. In addition, by replacing noisy diesel engines with quieter electric motors, noise pollution is also reduced, which enhances the quality of life in communities that live near large freight routes—another quantifiable, positive externality [13].

7.5 Comparative CBA: ERS vs. Alternative Decarbonization Pathways

ERS cannot be evaluated in isolation for its societal value. It must be judged against the main alternative ways of decarbonising the heavy-duty transport sector. To achieve this, a comparative CBA offers a suitable framework for the evaluation of the total system cost and net societal benefit of electrification for a major freight corridor under a variety of technological assumptions.

- **Scenario 1: ERS Network**: This scenario shows high initial infrastructure CapEx, but this is compensated by lower vehicle CapEx because of smaller batteries, and has the highest "well-to-wheel" energy efficiency of all options.

Table 7.2 Estimated TCO savings from battery downsizing for ERS-enabled HDVs

Vehicle & scenario	Required battery size (kWh)	Estimated battery cost (€)	Impact on vehicle purchase price (€)	Lifetime energy cost (€)	Lifetime maintenance cost (€)	Total cost of ownership (€/km)
40-ton HDV (static charging)	600–800	60,000–80,000	+150,000 (vs. Diesel)	200,000	80,000	0.85–1.05
40-ton HDV (ERS-enabled)	150–250	15,000–25,000	+80,000 (vs. Diesel)	185,000	80,000	0.70–0.90
Net TCO advantage	−75%	-€45,000	~€70,000 lower	~€15,000 lower	No Change	~ 15–20% lower

Note Costs are illustrative estimates for a long-haul truck over a 1-million-km lifetime. Battery cost assumes €100/kWh. Energy costs for ERS include usage fees. TCO figures are highly sensitive to electricity/diesel prices, subsidies, and route characteristics

Note Data synthesised from [1, 14, 15, 19–24]

- **Scenario 2: Megawatt Stationary Charging**: This pathway has a lower infrastructure CapEx per individual charging station than a continuous ERS would have. However, for long-haul operations to be seamless, a massive array of these high-power stations would have to be built across a broad area that would require major and expensive upgrades to the power grid [26]. Furthermore, this scenario requires very large and heavy on-board batteries (e.g. 1 MWh or more), which dramatically raise the purchase price of a car, reduce payload and lead to substantial operational costs associated with long hours of charging downtime and accelerated battery degradation [26].
- **Scenario 3: Hydrogen Fuel Cell Trucks**: This approach avoids the issue of battery size and charging time but requires the development of an entirely new and expensive infrastructure for hydrogen production, distribution, and refuelling [27]. The overall energy efficiency of the hydrogen pathway—from electricity generation to electrolysis, compression, transport, and conversion back to electricity in the vehicle—is significantly lower than that of direct electrification. Consequently, the cost of hydrogen fuel at the pump is currently several times higher than the equivalent cost of electricity, making it a less economically efficient solution at present [28].

By modelling the total societal costs and benefits of each scenario over a multi-decade analysis period, policymakers can make an informed decision about which pathway offers the most compelling long-term value proposition for public investment.

7.6 Financing and Investment Models for ERS Deployment

Transitioning from analytical justification to practical implementation involves solving the formidable challenge of funding and financing multi-billion-dollar ERS networks. The newness and high upfront costs of the technology require new models of financial organisation that combine leadership from the public sector with capital and expertise from the private sector.

7.6.1 The Role of Public Sector Funding

Given the high upfront capital costs, technology risks and the "network effect" problem (low value of the infrastructure until a critical mass of road and vehicles is achieved), large-scale ERS deployment is not seen to be feasible without significant public sector input, particularly at the early stages. This is a reality that is underlined by pilot projects all over the world. The eHighway trials in Germany are financed by the Federal Ministry of the Environment and the Federal Ministry of Transport.

Similarly, the Swedish ERS programme was initiated and financed by the government, via the Swedish Transport Administration (Trafikverket), which was the first ever ERS program in the world [7].

The recent cancellation of Sweden's planned permanent electric road project E20 is an important warning signal. The procurement was halted after private consortia's bid for the contract was far in excess of the government's allocated budget, showing the extreme dangers of cost uncertainty and the vulnerability of public funding commitments to new technologies. Public funding mechanisms are varied and may involve direct capital provision for infrastructure development, akin to the U.S. National Electric Vehicle Infrastructure (NEVI) and Charging and Fuelling Infrastructure (CFI) programs [29], and demand-side supports in the form of vehicle purchase subsidies to promote its adoption in fleets and form an initial user base [7].

7.6.2 Public–Private Partnership (PPP) Structures

PPPs, therefore provide a promising model of complex infrastructure delivery such as ERS. These models can make use of private sector capital, operating efficiency, and technological innovation—and the public sector can still exercise oversight and ensure that the project is consistent with public policy objectives. The crux and the key issue in a successful PPP is the right sharing of risks between the public and the private partners.

Several of the PPP models can be used for ERS applications. A Design-Build-Finance-Operate-Maintain (DBFOM) model, for example, brings the full life of the project under one private consortium. This entity, usually referred to as a concessionaire, would finance and build the ERS and operate and maintain it for a very long period of time (e.g. 30–50 years) in return for a predictable income stream. The latter gives rise to an intricate ecosystem of stakeholders, including the public authority (as the owner of the road), the infrastructure concessionaire, electricity suppliers, the technology providers, and the end-users (logistics companies). To encourage a healthy market, the roles need to be "unbundled"—the neutral infrastructure operator must be separated from the competitive electricity and mobility service providers to have equal access and pricing [30].

The repayment of the private investment may be in several ways. Under an availability model, the government makes regular payments to the concessionaire as long as the ERS infrastructure is operated according to pre-agreed performance standards. This model shifts construction and operational risk to the private sector but shifts the demand risk (i.e. the risk of low volumes) to the public sector. Conversely, in a concession-based model, the private operator gets the revenue directly from the users via toll, or fee (e.g. charged per kilometre driven or per kilowatt-hour consumed), thus shifting the operational and demand risk to the private partner. Models that are a hybrid between these two approaches are also common (Table 7.3).

Table 7.3 Risk allocation matrix for ERS PPP models

Risk category	Public procurement	Availability-payment PPP	Concession/toll-based PPP
Design & construction	Primarily public	Primarily private	Primarily private
Technology obsolescence	Primarily public	Shared	Primarily private
Demand/traffic volume	Public	Public	Primarily private
Operational performance	Public	Primarily private	Primarily private
Energy price volatility	Shared (pass-through)	Shared (pass-through)	Shared (pass-through)
Regulatory & political change	Shared	Shared	Shared

Note This matrix represents a typical allocation; specific risk-sharing is subject to negotiation in each PPP contract
Data synthesized from [7, 31–33]

7.6.3 Attracting Private Capital and Green Finance

To move ERS from pilot projects, mostly publicly funded, to bankable, scalable networks, projects must be designed to be "bankable". This requires them to attract the huge and exponentially expanding pool of private capital interested in sustainable and green investments, worldwide. ERS projects are well-suited to this form of financing, since they specifically contribute to climate-change mitigation and fit in with the intention of frameworks such as the EU Taxonomy for Sustainable Activities. This taxonomy is a typology that characterises economically sustainable activities, and helps in directing the investment capital to these activities by providing investors with more clarity and protection against greenwash [34].

In order to attract institutional investors, a strong pipeline of bankable projects has to be created based on clear and stable government policy, and with predictable, long-term returns. For an innovative technology like ERS, in which the risks are not yet regarded as low, multilateral development banks like the World Bank or the Asian Infrastructure Investment Bank (AIIB) can play a key role in de-rating the risks. Through financial innovations such as blended finance, first loss guarantee or concessional loans, these institutions can take up some of the initial risk of the project, which in turn "crowds in" private capital that otherwise would be more reluctant to invest. The financial model is only as good as the underlying legislation and regulatory framework. For private investors, legal certainty about the status of ERS infrastructure (is it part of the road or the energy grid?), the mechanisms for billing and revenue collection (e.g. calibrated per-kWh metering vs. flat-rate models), and rules relating to data privacy are certainly not mere technicalities but are fundamental to risk assessment and willingness to commit capital [7].

7.7 Strategic Recommendations for Financial and Policy Frameworks

The effectiveness of the implementation of Electric Road Systems (ERS) on the scale level depends on the presence of an integrated approach to technology, economy, and policy. Based on the above analysis, the following recommendations were put forward for policymakers and investors to deal with the complexity of ERS financing and implementation:

- **Adopt a Corridor-Based, Phased Deployment Strategy**: The economic and logistical scale of executing ERS, nationally and at a large scale, is not financially and logistically feasible as an en masse endeavour. A wiser course of action is to launch pilot and early commercial ventures on high-density freight routes where the business case is best. Connections between major seaports and hinterland logistics centres, for example, have high and predictable volumes of traffic and, as a result, maximise initial utilisation and thus demonstrate their economic viability. This channel-based approach allows the technical standards, operating process and finance schemes to be refined over time in a confined setting before being extended to a broader, more ambitious deployment. Such an approach would be congruent with strategic planning practice in Germany, which has identified a core network of 4000 km responsible for most long-haul fuel consumption, and with India's plans to identify high-density corridors for initial electrification efforts [35].
- **Prioritise and Mandate Standardisation**: The single non-financial threat to ERS scalability is market fragmentation caused by competing incompatible technologies. Lack of technical and commercial interoperability would result in infrastructure and vehicle operators' use of proprietary systems, which would inhibit competition and create a barrier to developing a seamless, cross-border network. Governments have to play a proactive role in promoting international cooperation in order to develop a standard for power transfer and communications, and billing systems. The process builds on existing research co-operations, for example, between Germany, France and Sweden. The requirement of one open standard will also create a competitive and innovative market of infrastructure and vehicle components, limiting price and increasing adoption [12].
- **Design Integrated "Whole-of-System" Policies**: ERS is a success that relies on a policy ecosystem that is coordinated to solve the infrastructure (supply) and vehicle (demand) side of the market simultaneously. A decentralised approach is virtual suicide. A good policy package should consist of:
- **Infrastructure Incentives**: Direct government funding, grants and tax credits like those under the U.S. Inflation Reduction Act are needed to overcome the first stage of capital barrier and to reduce the economic burden on early private investors [29].
- **Vehicle Incentives**: It is necessary to stimulate the demand for the infrastructure as well. This can be provided via subsidies, vouchers or tax measures for the purchase of ERS-compatible vehicles, thus allowing fleet operators to compensate

the high initial investment in vehicles and to initiate a critical mass of users since the beginning [7].

- **Regulatory Clarity**: Having a clear, stable and predictable regulatory environment is the most important to attract long-term private investment. Governments will need to provide clear rules on grid connection, electricity pricing and billing, as well as the legal status and ownership of the ERS assets, to give investors the confidence they need [30].
- **Leverage Blended Finance and Public–Private Partnerships for Scalability**: While public investment is crucial to get ERS projects started, private capital is needed to expand these projects into a network at the national level. Public funds should be used wisely to de-risk projects for private investors. Projects that are bankable, such as blended finance, where the initial losses are borne by the public or philanthropic capital or guaranteed by it, can be made attractive to institutional investors. The projects are to be designed in the form of long-term public–private approaches (PPP), where the risks are shared and the interests of all stakeholders are aligned. To tap into the ever-growing pool of sustainable investment funds globally, ERS projects must be designed to meet the international green finance requirements, including the EU Taxonomy for Sustainable Activities. It is a key strategy in moving to commercially independent infrastructure network pilot projects led by the government to a commercially self-sustained infrastructure network, as is projected in bold national-level programmes, including the National Infrastructure Pipeline in India [64].

References

1. Taylor & Francis Online (2025) Large-scale implementation of electric road systems: associated costs and the impact on CO_2 emissions. https://doi.org/10.1080/15568318.2019.1595227. Accessed 16 Oct 2025
2. Virta Global (2025) The future of electromobility—IEA's global EV outlook 2025. https://www.virta.global/blog/the-future-of-electromobility-ieas-global-ev-outlook. Accessed 16 Oct 2025
3. IFEU (2025) Opportunities for electric road systems in road freight economics. https://www.ifeu.de/fileadmin/uploads/Publikationen/Mobilit%C3%A4t/CollERS-2-Discussion-paper-3-Opportunities-for-electric-road-systems-in-road-freight-economics.pdf. Accessed 16 Oct 2025
4. Vectoral Journal (2025) Comparative study of conductive and inductive charging technologies for electric vehicles. https://vectoral.org/index.php/QJETI/article/download/90/83/178. Accessed 16 Oct 2025
5. MDPI (2025) Conductive electric road localization and related vehicle power supply systems. https://www.mdpi.com/2032-6653/13/1/22. Accessed 16 Oct 2025
6. Qmerit (2025) An EV fleet perspective of the total cost of ownership (TCO). https://qmerit.com/blog/electrification-of-transportation-and-the-total-cost-of-ownership-tco-a-fleet-perspective/. Accessed 16 Oct 2025
7. Trafikverket Bransch (2025) Business models and financing for the development of electric roads in Sweden. https://bransch.trafikverket.se/contentassets/19df180685ec467185af04e03f0cf03b/affarsmodeller-elvagar_slutrapport-180821-_en-final-190320.pdf. Accessed 16 Oct 2025

8. Energy Proceedings (2025) Technical and economic feasibility of electrified highways for sustainable mobility. https://www.energy-proceedings.org/wp-content/uploads/vol17/MITAB_2021_paper_122.pdf. Accessed 16 Oct 2025
9. DiVA Portal (2025) Electric road systems. https://www.diva-portal.org/smash/get/diva2:1046753/FULLTEXT01.pdf. Accessed 16 Oct 2025
10. PIARC (2025) Electric road systems: a solution for the future? https://www.piarc.org/en/order-library/29690-en-Electric%20road%20systems:%20a%20solution%20for%20the%20future. Accessed 16 Oct 2025
11. SINTEF (2025) Technology for dynamic on-road power transfer to electric vehicles. https://www.sintef.no/globalassets/project/elingo/18-0733-rapport-3-technology-for-dynamic-on-road-6-til-nett.pdf. Accessed 16 Oct 2025
12. Wikipedia (2025) Swedish transport administration electric road program. https://en.wikipedia.org/wiki/Swedish_Transport_Administration_electric_road_program. Accessed 16 Oct 2025
13. DiVA Portal (2025) Socioeconomic analysis of electric road systems. https://www.diva-portal.org/smash/get/diva2:1658492/FULLTEXT01.pdf. Accessed 16 Oct 2025
14. Global Fuel Economy Initiative (GFEI) (2025) Downsize EVs to accelerate electric mobility and diversify supply chains. https://www.globalfueleconomy.org/news/downsize-evs-to-accelerate-electric-mobility-and-diversify-supply-chains-says-gfei. Accessed 16 Oct 2025
15. MDPI (2025) Benefits of an electric road system for battery electric vehicles. https://www.mdpi.com/2032-6653/13/11/197. Accessed 16 Oct 2025
16. University of California, Davis (2025) Metrics & measurement methods. Electric vehicle research center. https://ev.ucdavis.edu/ssr-toolkit/metrics-and-measurement-methods. Accessed 16 Oct 2025
17. Rizon Truck (2025) EV total cost of ownership and why you can save money. https://www.rizontruck.com/blogarticle/ev-total-cost-of-ownership/. Accessed 16 Oct 2025
18. Clean Fleet Report (2025) The social and economic benefits of electric vehicle adoption in low-income communities. https://cleanfleetreport.com/the-social-and-economic-benefits-of-electric-vehicle-adoption-in-low-income-communities/. Accessed 16 Oct 2025
19. ResearchGate (2025) Total cost of ownership of electric vehicles: a synthesis of critical factors. https://www.researchgate.net/publication/394179700_Total_Cost_of_Ownership_of_Electric_Vehicles_A_Synthesis_of_Critical_Factors. Accessed 16 Oct 2025
20. ResearchGate (2025) The importance of powertrain downsizing in a benefit–cost analysis of vehicle lightweighting. https://www.researchgate.net/publication/315950851_The_Importance_of_Powertrain_Downsizing_in_a_Benefit-Cost_Analysis_of_Vehicle_Lightweighting. Accessed 16 Oct 2025
21. MDPI (2025) The economic feasibility of battery electric trucks: a review of the literature. https://www.mdpi.com/1996-1073/18/2/429. Accessed 16 Oct 2025
22. United States Environmental Protection Agency (EPA) (2025) Quantifying the emissions and health benefits of energy efficiency and renewable energy. https://www.epa.gov/sites/default/files/2018-07/documents/mbg_2-4_emissionshealthbenefits.pdf. Accessed 16 Oct 2025
23. United States Environmental Protection Agency (EPA) (2025) Quantifying energy savings and greenhouse gas (GHG) reductions. https://www.epa.gov/inflation-reduction-act/quantifying-energy-savings-and-greenhouse-gas-ghg-reductions. Accessed 16 Oct 2025
24. Greenhouse Gas Protocol (2025) Calculation tools and guidance. https://ghgprotocol.org/calculation-tools-and-guidance. Accessed 16 Oct 2025
25. Journal of Energy Systems (2025) Quantifying carbon emission reduction by adopting electric vehicles using a smart device. https://journal.esrgroups.org/jes/article/download/7204/4959/13228. Accessed 16 Oct 2025
26. Hydrogen Council (2025) Hydrogen insights—December 2023. https://hydrogencouncil.com/wp-content/uploads/2023/12/Hydrogen-Insights-Dec-2023-Update.pdf. Accessed 16 Oct 2025
27. Hydrogen Insight (2025) Average hydrogen fuel price at pump in the U.S. more than four times higher than in Japan and South Korea. https://www.hydrogeninsight.com/transport/average-hydrogen-fuel-price-at-pump-in-us-more-than-four-times-higher-than-in-japan-and-south-korea/2-1-1656000. Accessed 16 Oct 2025

28. HVTT Forum (2025) Swedish roadmap for electric road systems. https://hvttforum.org/wp-content/uploads/2019/11/Lindgren-SWEDISH-ROADMAP-FOR-ELECTRIC-ROAD-SYSTEMS.pdf. Accessed 16 Oct 2025
29. Congress.gov (2025) Public–private partnerships (P3s) in transportation. https://www.congress.gov/crs-product/R45010. Accessed 16 Oct 2025
30. IKEM (2025) Stakeholder model for the financing and billing of electric road systems (ERS). https://www.ikem.de/wp-content/uploads/2021/08/20210816_ERS-Akteursmodelle_EN.pdf. Accessed 16 Oct 2025
31. IKEM (2025) Stakeholder model for the financing and billing of electric road systems (ERS). https://www.ikem.de/en/publikation/stakeholder-model-for-the-financing-and-billing-of-electric-road-systems-ers/. Accessed 16 Oct 2025
32. CollERS 2 (2025) Business models, ownership, and financing strategies. https://electric-road-systems.eu/e-r-systems-wAssets/docs/publications/CollERS-1-Business-models-Ownership-and-Financing-Strategies.pdf. Accessed 16 Oct 2025
33. European Commission (2025) EU taxonomy for sustainable activities—finance. https://finance.ec.europa.eu/sustainable-finance/tools-and-standards/eu-taxonomy-sustainable-activities_en. Accessed 16 Oct 2025
34. OECD (2025) What to watch for at the 2025 OECD forum on green finance and investment. https://www.oecd.org/en/blogs/2025/10/what-to-watch-for-at-the-2025-oecd-forum-on-green-finance-and-investment.html. Accessed 16 Oct 2025
35. U.S. Department of Transportation (2025) Federal funding programs. https://www.transportation.gov/rural/ev/toolkit/ev-infrastructure-funding-and-financing/federal-funding-programs. Accessed 16 Oct 2025
36. Resources for the Future (2025) Operational versus capital expenditure risk in a clean energy transition. https://www.rff.org/documents/4393/Report_24-04.pdf. Accessed 16 Oct 2025
37. Stanford Law School (2025) Clean energy transportation services: a life-cycle cost framework. https://law.stanford.edu/wp-content/uploads/2020/06/20200520_LCXM_Full_Paper.pdf. Accessed 16 Oct 2025
38. Straits Research (2025) Electric highway (e-highway) market size, share & trends by 2033. https://straitsresearch.com/report/electric-highway-market. Accessed 16 Oct 2025
39. GOV.UK (2025) Transport notes. https://assets.publishing.service.gov.uk/media/57a08c72ed915d3cfd0013c8/C21-trn6.pdf. Accessed 16 Oct 2025
40. International Energy Agency (IEA) (2025) Batteries and secure energy transitions. https://iea.blob.core.windows.net/assets/cb39c1bf-d2b3-446d-8c35-aae6b1f3a4a0/BatteriesandSecureEnergyTransitions.pdf. Accessed 16 Oct 2025
41. MDPI (2025) Real energy efficiency of road vehicles. https://www.mdpi.com/1996-1073/18/8/1933. Accessed 16 Oct 2025
42. PubMed Central (PMC) (2025) Economy-wide impacts of road transport electrification in the EU. https://pmc.ncbi.nlm.nih.gov/articles/PMC9380419/. Accessed 16 Oct 2025
43. University of Chicago (2025) Calculating the cost of e-trucking. https://sing.uchicago.edu/2022/08/15/calculating-the-cost-of-etrucking/. Accessed 16 Oct 2025
44. Qmerit (2025) Cost analysis of EV home charging vs. public charging. https://qmerit.com/blog/comparing-long-term-cost-analysis-of-ev-home-charging-vs-public-charging/. Accessed 16 Oct 2025
45. University of Rhode Island (2025) Quantifying the impact of traffic on electric vehicle efficiency. https://digitalcommons.uri.edu/cgi/viewcontent.cgi?article=2113&context=mcise_facpubs. Accessed 16 Oct 2025
46. U.S. Department of Energy (2025) Hydrogen and fuel cells program record 18003: current status of hydrogen delivery and dispensing costs. https://www.hydrogen.energy.gov/docs/hydrogenprogramlibraries/pdfs/18003_current_status_hydrogen_delivery_dispensing_costs.pdf?Status=Master. Accessed 16 Oct 2025
47. U.S. Department of Energy (2025) Hydrogen fueling station cost analysis. https://www.hydrogen.energy.gov/docs/hydrogenprogramlibraries/pdfs/21002-hydrogen-fueling-station-cost.pdf?Status=Master#:~:text=Table%202%20summarizes%20the%20costs,%2Fday%20per%20dispenser7). Accessed 16 Oct 2025

48. Siemens (2025) Third eHighway field trial starts in Germany. https://press.siemens.com/global/en/pressrelease/test-field-ehighway-project-ewaybw. Accessed 16 Oct 2025
49. Continental AG (2025) New partnership: continental and siemens mobility to supply trucks across europe with electricity from overhead lines. https://www.continental.com/en/press/press-releases/20210729-ces-siemens-ehighway/. Accessed 16 Oct 2025
50. Trafikverket Bransch (2025) Electric road E20, Hallsberg–Örebro. https://bransch.trafikverket.se/en/startpage/projects/Road-construction-projects/electric-road-e20-hallsbergorebro/. Accessed 16 Oct 2025
51. PPP Knowledge Lab (2025) Public–private partnership agency. https://pppagency.gov.ua/knowledge-lab/. Accessed 16 Oct 2025
52. League of American Bicyclists (2025) Public–private partnerships for transportation. https://bikeleague.org/wp-content/uploads/2023/03/PublicPrivatePartnerships_September2014.pdf. Accessed 16 Oct 2025
53. Principles for Responsible Investment (PRI) (2025) EU sustainable finance taxonomy. https://www.unpri.org/policy/eu-policy/eu-taxonomy. Accessed 16 Oct 2025
54. Norton Rose Fulbright (2025) EU taxonomy for shipping: a transition to sustainability. https://www.nortonrosefulbright.com/en/knowledge/publications/92023ad9/eu-taxonomy-for-shipping-a-transition-to-sustainability. Accessed 16 Oct 2025
55. OECD (2025) Finance and investment for climate goals. https://www.oecd.org/en/topics/finance-and-investment-for-climate-goals.html. Accessed 16 Oct 2025
56. OECD (2025) Climate finance and the USD 100 billion goal. https://www.oecd.org/en/topics/climate-finance-and-the-usd-100-billion-goal.html. Accessed 16 Oct 2025
57. OECD (2025) Global outlook on financing for sustainable development 2025. https://www.oecd.org/en/publications/global-outlook-on-financing-for-sustainable-development-2025_753d5368-en.html. Accessed 16 Oct 2025
58. World Bank PPP (2025) Knowledge lab. Public–private partnership. https://ppp.worldbank.org/public-private-partnership/taxonomy/term/1364?page=20. Accessed 16 Oct 2025
59. NITI Aayog (2025) Electric vehicles in India. https://niti.gov.in/sites/default/files/2025-08/Electric-Vehicles-WEB-LOW-Report.pdf. Accessed 16 Oct 2025
60. Fraunhofer Institute (2025) The state of electric road systems. https://publica.fraunhofer.de/bitstreams/6a66b7bc-8e3e-48a0-a276-f9d4a6ee4dd1/download. Accessed 16 Oct 2025
61. ICRA Limited (2025) National infrastructure pipeline (NIP)—research summary report (ID: 6216). https://www.icra.in/Rating/DownloadResearchSummaryReport?id=6216. Accessed 16 Oct 2025
62. ICRA Limited (2025) National infrastructure pipeline (NIP)—research summary report (ID: 5580). https://www.icra.in/Rating/DownloadResearchSummaryReport?id=5580. Accessed 16 Oct 2025

Chapter 8
Integration of Electric Road Systems with Public Transport and Freight Networks

8.1 Introduction

Electric Road Systems (ERS) are a category of technologies that provide electrical power to vehicles during motion, which essentially changes the paradigm of transport electrification on a fundamental level through the provision of electricity supply infrastructure, the road itself, the power transfer system, the road operation system and the transport vehicles appropriately equipped. Three main ERS technologies are being demonstrated and tested globally, which include conductive overhead lines (catenary systems), conductive rails on the ground or integrated in the road surface, and wireless (inductive) power transfer through coils beneath the asphalt [1]. The combination of these technologies in existing transport networks and future transport networks offers a radical route towards decarbonising high-utilisation vehicle platforms, like passenger buses and heavy-duty freight trucks.

The strategic necessity of moving, moving charge comes out of the nature of operations and economics of mobility of commercial transport in the face of the static model of commercial transport charging. Whereas overnight depot charging and high-power opportunity charging can work well with most electric vehicles, they create high costs to fleets based on high asset utilisation, and require large on-board batteries to enable a vehicle to travel long distances, as well as the large-scale capital and weight cost of large on-board batteries [2]. ERS can solve this by changing the paradigm of seeing vehicles as the primary energy storage tool to seeing infrastructure as the energy storage tool. ERS enables vehicles to run continuously along major routes with considerably smaller batteries that are somewhat used as a buffer when the vehicle travels outside the network, passes passing manoeuvres or fills in gaps within the electrified network. This method is a good way to extend the range, which reduces range anxiety and brings the performance profile of electric vehicles closer to the conventional one.

This chapter shows how ERS can revolutionise the way people transport and freight logistics in the urban and intercity areas. The analysis will begin with the

R. Agnihotri, *Electrified Road Systems (ERS) for Emerging Asia*, Springer Tracts on Transportation and Traffic 24, https://doi.org/10.1007/978-3-032-16927-3_8

discussion of ERS implementation in the public transport industry, with the emphasis put on electric bus corridors and Bus Rapid Transit (BRT). It then moves to the ERS application to the urban and intercity freight, where operational models of heavy-duty trucks and logistics fleets are analysed. The following sections give a more specific description of the technical conditions of vehicle and infrastructure adaptation, a quantitative evaluation of the benefits of operation and economy, and a discussion of the interdependence with urban planning and smart-city initiatives. An assessment of world experience in the form of a comparison of case studies offers a context to a proposed integration framework in accordance with the peculiar conditions of the emerging economies in Asia. The chapter is concluded by discussing the major obstacles to large-scale implementation, namely, technical, economic, and institutional ones, and making a future prospectus where ERS is coupled with digital twins and Vehicle-to-Grid (V2G) technology, thus preparing the discussion on smart-city integration in Chap. 9.

8.2 ERS Integration in Public Transport

The adoption of Electric Road Systems into the public transport systems, especially in the high-frequency city bus amenity, is an intriguing approach to the challenge of electrifying the fleet of buses. In services like Bus Rapid Transit (BRT), where dedicated or priority lanes are common, ERS can be implemented with little additional interference to regular traffic, to create a highly efficient, zero-emission mass-transit corridor [3]. This synergy will effectively convert the bus-as-tram concept of BRT into an electric tram on tyres, with the sustained power of light rail and the freedom of movement of a bus, which can drop out of the electrified corridor to serve the peripheral regions on its small on-board battery.

8.2.1 Operational Advantages: Continuous Charging and Fleet Optimisation

The main operational benefit of ERS to the operation of the public transport is that it is able to charge continuously whilst in motion. This essentially changes the trade-off between battery size, vehicle range, and charging downtime, which limits conventional battery-electric bus (BEB) operations. Under ERS, the battery can be considerably smaller and lighter; since the battery is not the major energy source, but a backup, the battery will be smaller in size. Technical experimental projects and pilot projects indicate that battery sizes can be reduced by 50–80%. Such a decrease triggers a virtuous cycle.

1. **Reduced Vehicle Weight**: The lighter the battery, the less the total vehicle weight, resulting in less energy used per kilometre.

2. **Increased Capacity**: The space and weight saved can be re-allocated towards more passengers in the vehicle, hence improving the economic efficiency of the individual vehicles.
3. **Lower Capital Cost**: The cost of batteries constitutes a large percentage of a BEB's initial cost. The small size of the batteries has a direct impact on the capital expenditure of the vehicle.
4. **Enhanced Fleet Uptime**: ERS facilitates the least out-of-service time with every bus since it no longer requires long depot charging or intermediate stops at fast chargers. This greater use of assets allows a transit authority to travel within its schedule using fewer vehicles than in a depot-charging model, and this provides significant savings in the general purchase and maintenance costs of the entire fleet.

8.2.2 Infrastructure Considerations and System Design

The deployment of ERS to a bus corridor needs thorough planning and consideration of the existing urban infrastructure. In the case of in-ground technologies (conductive rail or inductive), this can be achieved by retrofitting dedicated bus lanes that can require trenching of the pavement to install the charging equipment. Even though it is aesthetically inoffensive, this method means that road maintenance and resurfacing programs have to be coordinated with (thus creating visual effects, i.e. masts and overhead cables) that cannot fit historically or aesthetically sensitive urban environments.

Regardless of the chosen technology, ERS implementation requires a good electrical infrastructure, such as substations at strategic points along the route to interface with the local grid and convert electricity to the correct voltage and current to use with the system [3]. Terminal stations and layover points may be fitted with stationary charge pads of ERS (conductive or inductive) to optimise the state of charge of the vehicle during scheduled breaks. Figure 8.1 shows the design of an ERS-enabled BRT corridor

illustrating key components including the dedicated lane with embedded charging infrastructure, roadside substation, and terminal charging facilities.

8.2.3 Case Example: Evolution Road in Lund, Sweden

A ground-level electric rail system (ERS) conductor is an innovative example of conductive systems, ground-level electric rails that are used to transport buses publicly, with Evolution Road in Lund, Sweden, being such a project. The project was commissioned by the Swedish Transport Administration and involved the installation of a one-kilometre conductive rail in a special bus lane along a pedestrian street, Getingevägen. Solaris electric bus was adapted to the current collector so as to test

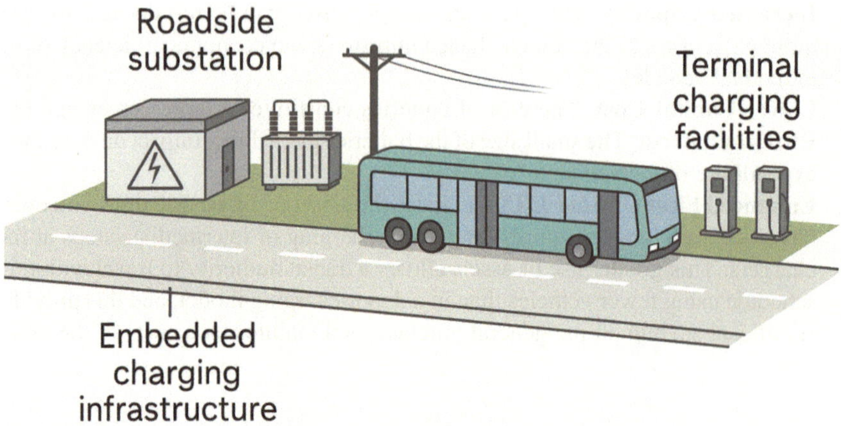

Fig. 8.1 Schematic layout of an ERS-enabled bus rapid transit (BRT) corridor

the system in the real operating conditions. In this technology, which Elonroad developed, a bus can be charged dynamically as it travels, and when it is not moving, it can be charged statically.

One of the most imperative operations metrics as a result of the project is that charging a kilometre on the electric road will provide the bus with enough power to move a distance of about three kilometres, hence making the system credible as a range extender. The empirical knowledge, which can be gained as a result of the endeavour, will be used to develop future modalities of transport without fossil fuel, and one of the key benefits is that battery capacities may be reduced by 5080 per cent, which in turn will lower both expenses and environmental consequences.

The safety architecture of the system is focused on the mitigation of risks; the rail will be energised only at short sections right below a licensed vehicle and will not introduce any risk to pedestrians and other road users. Where the Paris Bus 2025 programme has focused on depot charging, new endeavours like the Tzen 4 BRT line are considering high-power static ground charging, and a pilot on the A10 highway should consider dynamic bus wireless charging. These changes show an increased openness to the ERS-type solutions in large European cities.

The comparative advantages of ERS over traditional charging methods for bus fleets are summarised in Table 8.1, highlighting the fundamental trade-offs between infrastructure investment and operational efficiency.

Table 8.1 Comparative metrics for ERS vs. depot and opportunity charging for electric bus fleets

Metric	Depot charging (overnight)	Opportunity charging (en route)	ERS (in-motion charging)
Battery size (capacity)	Very Large (e.g., 300–500 kWh)	Medium (e.g., 150–300 kWh)	Small (e.g., 50–100 kWh)
Battery cost & weight	High	Moderate	Low
Charging downtime	4–8 h (overnight, non-operational)	5–15 min per charge (multiple times per day, operational)	Minimal to none (charging during operation)
Fleet uptime/asset utilization	Lower (vehicle unavailable during charging)	Moderate (short disruptions to schedule)	Highest (near-continuous availability)
Required fleet size	Largest (extra buses needed to cover for charging units)	Larger	Smallest
Infrastructure cost	Moderate (depot chargers and grid upgrades)	High (multiple high-power chargers along routes)	Very High (cost per km of electrified road)
Grid impact	Concentrated at depots, can be managed for off-peak hours	High-power peaks at multiple points during the day	Distributed load along the corridor, more stable demand

Sources [4]

8.3 Integration with Urban and Intercity Freight Networks

The road freight industry, namely long-haul heavy-trucking, presents a serious obstacle to decarbonisation as it has a high energy requirement, a large geographical scope of operation, and is sensitive to the weight of the cargo. Battery-electric trucks need a large, heavy and costly energy storage system, which has the effect of reducing cargo capacity and necessitating long charging intervals which undermine operational feasibility. ERS is a more practical and scalable option by providing the continuity of power over key freight routes, thereby connecting other key economic sectors, including ports, manufacturing hubs, and logistics centres.

8.3.1 The Hybrid Operational Model: A Transitional Framework

One of the outstanding aspects of ERS integration in freight transport is the use of a hybrid framework of operation. ERS trucks do not necessarily depend on the electrified infrastructure, but are instead commonly designed as hybrids, which means adding ERS-receiving power infrastructure as a secondary energy source, which may be a smaller battery pack or a traditional diesel/biodiesel engine. This architecture provides the flexible operation required in vehicles. Electric trucks can be attracted when passing through electrified highway spines and can be easily transferred to the electric source on board when passing through non-electrified secondary roads (ribs) or during the first and last miles of a trip between a logistics hub and a final destination. The Spine and Ribs model provides a practical, cost-effective approach to phased implementation to enable governments and other investors to maximise the utility of initial investment in infrastructure by focusing on the busiest freight lanes [5].

Case Example: eHighway Germany (Catenary System)

The eHighway programme in Germany is the largest demonstration of catenary-based ERS in heavy freight, in the world. There are some pilot projects, the most famous being the ELISA (Electrified, Innovative Heavy Freight on the Autobahn) test on a 10-km stretch of the A5 autobahn between Frankfurt and Darmstadt. The engineered system by Siemens Mobility makes use of two overhead power lines to power specifically designed Scania hybrid trucks with 670 V DC [6].

The vehicle-side technology focuses on a smart pantograph that is capable of automatically connecting and disconnecting with the overhead lines at up to 90 km/h/ to allow trucks to travel in and out of the electrified lane without stopping. The system has a grid-to-wheel efficiency of more than 80 at a percentage, approximately twice that of a standard internal-combustion engine. In its long-term vision, the German government plans to electrify at least 4,000 km of the autobahn network by 2030, indicating that it is willing to make a long-term investment in this technology as a pillar of freight decarbonisation. The selection of a catenary system demonstrates a conscious emphasis on a theoretical solution to the commercial freight market because this technology best serves high-sided vehicles (trucks and buses). Figure 8.2 shows a schematic map of a freight corridor based on the catenary.

Case Example: Smart Road Gotland (Inductive System)

Unlike the German focus on conductive catenary systems, the operational feasibility of wireless inductive electric road systems (ERS), both in freight and in public transportation, was demonstrated in the Smart Road Gotland project in Sweden. The project involved installing 1.65 km of inductive charging coils on the ground under a public roadway between the airport and city centre in Visby under the leadership of Electreon. This infrastructure made it possible to dynamically charge a 40-tonne electric truck and an electric bus, thus depicting the prospects of a mutual and universal charging platform that can be used by various types of vehicles. The main findings, which were obtained due to the conclusion of the project in 2023, were related to the technical maturity and robustness of the system. The truck was capable of undergoing a power transfer of 100 kw whilst at a speed of 80 km/h. In addition, the technology proved tough in the brutal winter Nordic conditions, where it could work in snow and ice at temperatures as low as 23 °C, and it showed no decrease in charging ability. The success of the project emphasises the fact that in-ground systems provide a more aesthetically friendly alternative to catenaries, which may

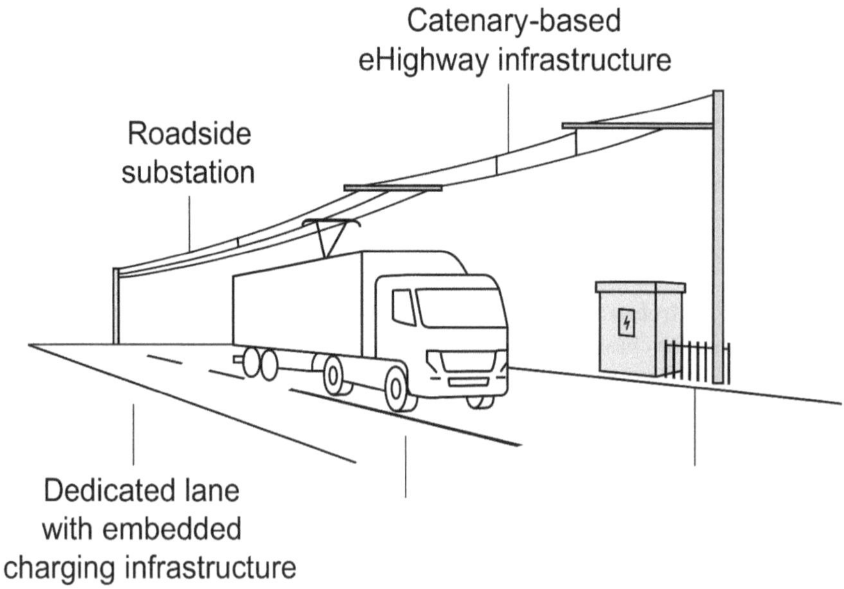

Fig. 8.2 Schematic of a freight corridor with a catenary-based electric road system (eHighway)

be more suitable in routes going through scenic or urban locations. Nonetheless, this universality also adds more complexity to the process of accessing and billing a wide range of users, which is a specific strategic and business model option in comparison to the dedicated user approach demonstrated by catenary systems.

8.4 Technical Requirements for Integration

An efficient implementation of the ERS into transport systems requires coordinated changes both at the vehicle level and in the road networks reinforced with efficient fleet management systems. The particular technical aspects would greatly differ depending on the chosen ERS technology: catenary, rail, or inductive.

Vehicle-Side Modifications

To utilise an ERS, vehicles must be equipped with specialised power transfer equipment and onboard power electronics.

- **Conductive (Catenary)**: buses and trucks are supposed to have an intelligent, roof-mounted pantograph. This gadget is designed in a manner that it is able to automatically lift and link to the power overhead lines and disengage when exiting the electrified lane or in passing the overtaking moves, all without the vehicle halting at highway speed. The system has sensors to identify the power lines and control the connection process. An onboard power converter is also necessary to convert high-voltage DC grid power (e.g., 670 V DC in eHighway system) to the power needed by the battery and electric motor in the vehicle.

- **Conductive (Rail)**: Automobiles using ground rails should have an underbody current collector, also known as a pickup. This mechanical arm is lowered to provide physical contact to the electrified rail in the road surface and slides down it to pass power [3]. This design needs to be strong enough to accommodate road debris as well as changing weather conditions and hold a constant electrical contact.
- **Inductive (Wireless)**: In the case of wireless charging, the vehicle has a receiver coil (pad) on the chassis, one or more. To enable power transfer through magnetic resonance, these coils are placed in alignment with transmitter coils that are installed within the road to provide power to the high-frequency AC power induction in the receiver coil to create DC power to charge the battery and power the motor [7]. The system also needs power electronics to convert high-frequency AC power in the receiver coil into DC power to charge the battery and power the motor.

In any case, an advanced onboard energy management system must be in place to regulate the flow of electricity between the ERS interface and the battery, and traction system, and the switching of the on and off network power.

8.4.1 Road Infrastructure Adaptation

Adapting road infrastructure for ERS is a significant civil and electrical engineering undertaking.

- **Catenary Systems**: To support overhead contact and messenger wires, the installation of roadside masts and foundations in regular intervals (usually at intervals not exceeding 60 m) is required. Although this method produces little direct effect on the actual road pavement, it presents a conspicuous visual feature and would require keen planning of the roads around bridges, tunnels and gantries.
- **In-Ground Systems (Rail and Inductive)**: Such systems are more intrusive to the road structure. Installation can also be done by trenching the asphalt to a depth of several centimetres to put either conductive rails or segments of inductive coils in place, which can impact the structural integrity of the road as well as complicate maintenance operations, as resurfacing of roads becomes difficult.
- **Electrical Supply**: Any ERS technologies need a network of roadside substations to be connected to the medium-voltage electrical grid (e.g., 1030 kV) and convert the power to proceed along the ERS segments [3], which are a part and parcel of the infrastructure.

8.4.2 Fleet Management and Telematics

The high performance of the ERS-enabled fleet is based on the sophisticated digital control equipment. The standard fleet management software will need to be extended to include ERS-specific features [8]. The telematics systems will be important to monitor the real-time position of any vehicle, speed, and battery state-of-charge, which will be transmitted to a central managerial platform and used to perform dynamic scheduling, vehicle routes, and charging plans to optimise the utilisation of the ERS infrastructure availability. Moreover, such systems are also necessary in terms of access control and billing, and they can identify authorised vehicles via wireless communication and meter energy use and conduct payments automatically, as shown in the eRoadArlanda project and the Smart Road Gotland project [9].

One of the major difficulties with scaling ERS is the absence of interoperability. The systems of various vendors currently cannot be compatible, such that one vehicle that is fitted with a specific system cannot be used with another system. This poses a threat of vendor lock-in and incomplete networks, which prevents the creation of a smooth, cross-border ERS. Scalability, especially of international freight, will require the introduction of industry-wide technical standards on physical interfaces, communications protocols and voltages. This standardisation is arguably the most important technical challenge to be surmounted in order to make ERS successful in the long term. Table 8.2 summarises the technical compatibility of the various types of vehicles with the main ERS technologies.

8.5 Operational and Economic Benefits

In the implementation of Electric Road Systems (ERS), there are a series of operational and economic benefits that make the overall cost of ownership (TCO) of electric fleets significantly lower. The economic ERS rationalisation does not just lie in a naive payback calculation of the infrastructural investment; the basis is a system-level analysis that sums up cost savings realised on the whole fleet of vehicles throughout its operation period [11].

8.5.1 Reduction in Fleet Battery Size and Cost

The most apparent and direct economic advantage that comes with ERS is the ability to significantly reduce the size of onboard battery assemblies. ERS eliminates the need to have batteries that have to support all-day-long functions on one charge. The pilot testing and theoretical analysis of the empirical results show that battery capacity can be reduced by half or two-thirds in ERS-enabled automobiles compared to similar cars that use only batteries. This decline creates several far-reaching economic effects.

Table 8.2 Technical compatibility matrix for ERS technologies and vehicle types

Vehicle type	Conductive overhead (catenary)	Conductive ground-level (rail)	Wireless inductive
City bus (12 m)	**High suitability**: Proven technology (trolleybus heritage); high power transfer	**High suitability**: Proven in pilots (e.g., Lund); low visual impact	**High suitability**: Proven in pilots (e.g., South Korea); no mechanical wear
Articulated bus (BRT)	**High suitability**: Ideal for high-capacity, dedicated-lane transit	**High suitability**: Well-suited for urban BRT corridors	**Feasible**: Suitable for BRT; may require multiple receivers for higher power
Light commercial van	**Not applicable**: Vehicle height is insufficient to reach overhead lines	**Feasible**: Technically possible; economic case depends on utilization	**Feasible**: Technically possible; well-suited for last-mile delivery fleets
Heavy-duty truck (40t)	**High suitability**: Most mature high-power solution; proven in German pilots	**Feasible**: Technically possible; requires robust pickup for high speeds/loads	**Feasible**: Demonstrated in pilots (e.g., Gotland); power levels are increasing
Passenger car	**Not applicable**: Vehicle height and power requirements are incompatible	**Feasible**: Technically possible; competes with mature static charging market	**Feasible**: Technically possible; main focus of some developers for mass-market

Sources [10]

- **Lower Vehicle CapEx**: Batteries are the most expensive element of an electric vehicle with a high power consumption; the smaller they are, the less a single bus or truck will cost to buy, and the more profitable conversion to a fleet will be.
- **Reduced Weight and Increased Payloadb**Smaller batteries will help reduce curb weight, which, in addition to decreasing energy consumption (reduced kWh/km), will increase the potential payload capacity and, most importantly, the earnings each truck can make.
- **Lower Replacement Costs**: Vehicle batteries are limited in terms of life cycle and are associated with expensive replacement. Smaller batteries under ERS control that are allowed to operate within an optimal range of state-of-charge that does not allow deep discharges had a higher service life, hence reduced lifecycle replacement costs.

8.5.2 Reduced Charging Downtime and Improved Asset Utilisation

Time is capital to commercial operators. ERS helps to eradicate non-productive flat rates of charge since the vehicles can recharge as they move. The mechanism brings out a dramatic increase in the utilisation of assets. A vehicle fleet fitted with ERS can work longer hours in a day as compared to the diesel equivalent, so that those operators are able to meet service or delivery timelines with fewer vehicles. The reduced fleet size will represent a serious saving in capital expenditure, and can frequently offset a sizeable proportion of the ERS infrastructure cost.

8.5.3 Lower Energy and Maintenance Expenditure

Direct electrification made possible through ERS is an extremely efficient method of propulsion. To take an example of that, catenary systems are capable of grid-to-wheel efficiencies of 80–85% compared to the approximate 40% of diesel engines. This, combined with the fact that the cost of electricity is usually lower and more predictable compared to the cost of diesel fuel is translates into significantly reduced energy cost per kilometre. In the case of freight, it has the implication of a better cost per ton-kilometre. Additionally, the less dependency on batteries and simpler mechanics of the electric powertrain, in comparison to internal combustion engines, provide a lower cost of maintenance during the lifetime of the vehicle. A study conducted in Germany predicted that using an eHighway would save EUR 16,000 of the fuel used by a 40-ton truck over 100,000 km.

8.5.4 Quantified Emission Reductions

The first advantage of the ERS to the environment is that almost no tailpipe emissions, such as carbon dioxide, nitrogen oxides and particulate matter (PM10) are emitted. The overall climate effect depends on the carbon intensity of the electric grid, but when the electricity grid is supplied by renewable sources, ERS will enable truly zero-emission transport. In a study commissioned by the French government, it was concluded that ERS would lead to a reduction in road freight emissions of Carbon Dioxide by up to 86% compared to diesel. The largest societal benefits of an inductive ERS corridor in Spain were found to be associated with positive effects on the environment (GHG decreases) and minimised vehicle operating expenses (Coste et al. 2020). Figure 8.3 gives a comparative visualisation of the cost saving and emission saving bestowed by ERS against alternative powertrain technologies.

8.6 Urban Planning and Infrastructure Integration

Electric Road Systems (ERS) does not just stop at the normal scope of transport engineering; it is more of a strategic urban planning project that requires extensive coordination with city master plans, intelligent infrastructure, and institutional cooperation. ERS should be viewed as a foundational layer of urban mobility since it can

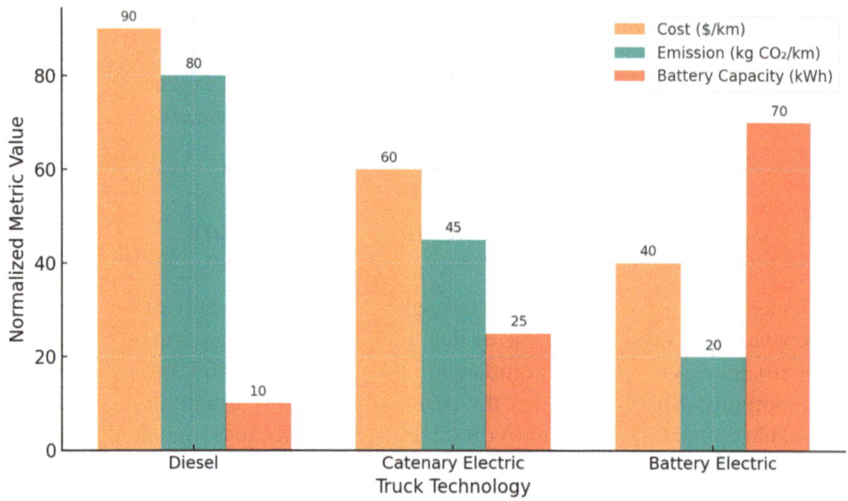

Fig. 8.3 Comparative cost, emission, and battery capacity metrics for different heavy-duty truck technologies

be used as a structural platform on which new development will be built in order to direct the evolution of cities towards more efficient and sustainable forms.

8.6.1 Integration with Smart City Systems

A digital system is inherently interconnected and networked, making an ERS a natural part of the ecosystem of a smart city [12]. The system constantly produces an abundance of real-time information about traffic movement, vehicle type, energy use, and grid demand. This data layer is a good asset that may be implemented in other city management platforms.

- **IoT-Based Traffic Management**: ERS information may provide real-time, detailed information on traffic jams and speed along arterial routes. This data can be relayed to a central traffic control centre of a city to decide how best to optimise the timing of the signals, control traffic movement, and offer real-time routing information to traffic drivers.
- **Smart Tolling and Automated Payments**: This can be combined with automated tolling systems, which would make it appear as a seamless experience of paying-as-you-go with energy and road usage, both [13].
- **Grid Management**: ERS provides utilities with better grid load management by providing predictable and real-time information on the energy needs of the transport sector. This will be critical in keeping stability, especially as the ratio of variable renewable energy sources increases.

8.6.2 ERS in City Master Plans and Transit-Oriented Development (TOD)

Urban planning requires proactive planning, without which the benefits of ERS cannot be maximised. Planners can steer development and investment by establishing a strong incentive towards the location of logistics centres in the vicinity of freight ERS corridors and high-density residential and commercial development in the vicinity of public-transport ERS corridors by designating key freight and public-transport routes as prospective ERS corridors in the long-term city master plans.

The strategy has a high level of synergy with the principles of Transit-Oriented Development (TOD). TOD aims to create small, pedestrian, mixed-use neighbourhoods around high-quality public-transport stations [14]. ERS can provide the technological base of the high-capacity, high-frequency transport (including BRT) that a successful TOD cannot do without. On the other hand, the concentration of people and employment in the TOD provides the stable ridership that will support the heavy capital investment in ERS infrastructure, and thus creates a reciprocal relationship between land use and transport technology.

8.6.3 Collaboration Between City Transport Authorities and Utilities

ERS is a multi-sectoral project that is difficult to deploy by a single organisation. It needs a powerful system of cooperation between various public and non-governmental stakeholders. Key actors include.

- **City Transport Authorities**: Tasked with the planning of road networks, right-of-way management and the incorporation of ERS into the comprehensive urban mobility plan.
- **Public Transit Operators**: The major customers of city ERS, charged with the responsibility of fleet acquisition, operation and planning.
- **Utility Companies:** Tasked with the role of determining the capacity of the grid, planning and implementing grid upgrades and setting tariff regimes of the electricity provided to the ERS.
- **Urban Planning Departments**: Planning these corridors into the long-term land-use plans and development regulations.

The requirement to have a joint governance structure, like the use of a special task force, to facilitate these activities and create alignment with the objectives can ensure successful implementation, starting with the initial planning, construction and operation. As shown in Fig. 8.4, ERS is interconnected in an urban system.

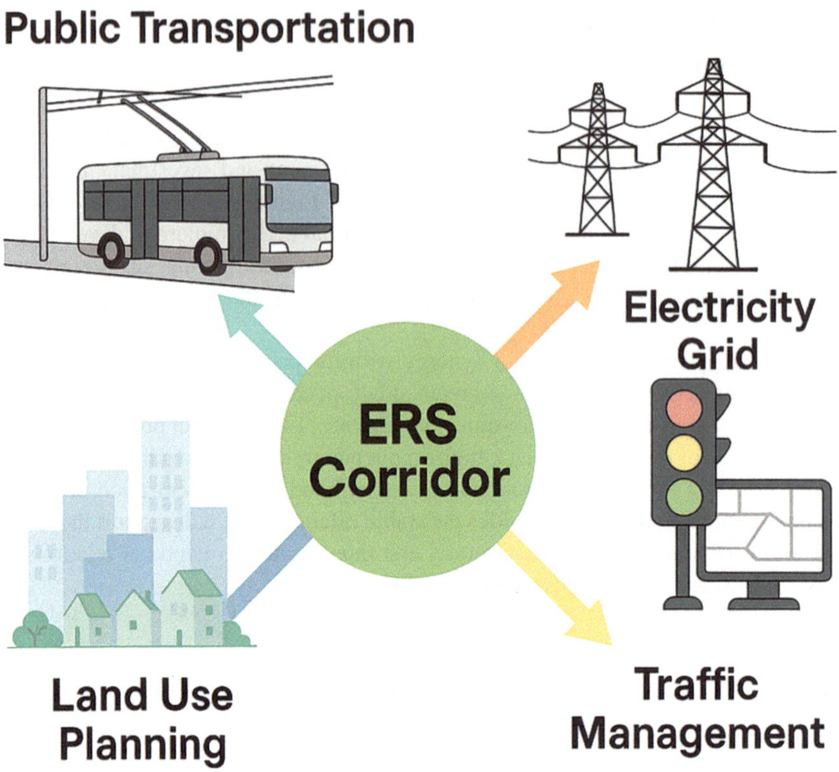

Fig. 8.4 Urban integration diagram illustrating the synergy between an ERS corridor and key city systems

8.7 Case Studies and Global Best Practices

The world has been able to use demonstration and pilot projects to enhance the technology of ERS between the concept and operational reality. Such tests have offered priceless information on technical performance, economic feasibility and implementational challenges. When a comparative analysis is performed, one can see that there are new tendencies in the strategies of the implementation of ERS by different regions, based on their own industrial potential and priorities of their policies and the geographic location.

- **Sweden**: Sweden has been the first to introduce a two-tracking method of testing in-ground technologies. The Evolution Road project at Lund demonstrated that a conductive rail system of mass-duty trucks and an urban bus can be charged along the same railway infrastructure, and the Smart Road Gotland project demonstrated that a wireless inductive system can charge a heavy truck and a bus simultaneously, both in a live traffic environment, demonstrating its compatibility with the mixed-traffic setting. The projects make Sweden a pioneer in creating ERS solutions that are less visible in the visual features of the urban and semi-urban environment. Yet, the latest move towards halting procurement of a permanent ERS on the E20 highway because of the cost more than the budget amounts to huge financial challenges to large-scale implementation.

- **Germany**: Germany has strongly concentrated on catenary systems to decarbonise its key long-haul freight industry. The best use of overhead conductive technology has been demonstrated by the eHighway programme with major test tracks on the A5 (Hesse), A1 (Schleswig–Holstein) and B462 (Baden-Wurttemberg) federal highways, which were developed in the context of a strong and successful public–private partnership that involved Siemens, truck manufacturers including Scania and logistics companies. The German strategy is an approach based on targeted use, and the priority given to the highest-impact use case, heavy freight on the autobahn network.
- **South Korea**: South Korea was one of the first, with its Online Electric Vehicle (OLEV) technology, which is a dynamic wireless charging model, being the first commercial bus service in the world. In Gumi city and Sejong city, bus services began operation in 2013, although these programs were trailblazers in showing how inductive ERS could be used in commercial practice, some of the lines were later closed because of the problems with ageing infrastructure. The experience is a lesson that long-term planning on maintenance, durability and lifecycle costs in ERS projects is critical.
- **United States**: The U.S. approach to freight electrification has largely centred on incentives for battery-electric trucks and the deployment of static fast-charging infrastructure, particularly in California. However, a notable ERS pilot was conducted at the Ports of Los Angeles and Long Beach, where Siemens and Volvo collaborated on a catenary system to power hybrid trucks moving cargo containers, demonstrating the technology's potential for reducing emissions in heavily polluted port environments.

This global overview shows a clear trend moving from initial questions of "technical feasibility" toward more complex assessments of "economic and operational viability." Early projects proved the technology's work. Current and future efforts are focused on refining business models, establishing standards, and proving that ERS can be deployed, operated, and cost-effectively maintained at scale. Table 8.3 provides a comparative summary of these key international projects.

8.8 Integration Framework for Emerging Asia

This is a twofold challenge that the emerging economies of Asia, especially India, Thailand, Indonesia, and Bangladesh, are facing to ensure they absorb significant projected increases in the freight and passenger transportation demands, on the one

Table 8.3 Comparative analysis of global ERS case studies

Country	Project name	Sector	Technology	Coverage length (km)	Power transfer (kW)	Key impact/finding	Scalability status
Sweden	EVolutionRoad (Lund)	Public transport	Conductive rail	1	150–200	Proved viability for urban buses in live traffic; 1 km charge enables 3 km drive	Pilot completed; informed national strategy
Sweden	Smartroad Gotland	Freight & public transport	Wireless inductive	1.65	100 (for truck)	Demonstrated shared use and durability in harsh weather; high user satisfaction	Pilot completed; technology being commercialized globally
Germany	eHighway (A5, ELISA)	Freight	Conductive overhead	10 (5 km each way)	~600 (per truck)	Confirmed high efficiency (>80%) and operational reliability at highway speeds	Field trial ongoing; informing national scale-up plan (4,000 km)
South Korea	OLEV (Gumi)	Public transport	Wireless inductive	~12 (route length)	52–85	First commercial deployment of dynamic wireless charging for buses	Some lines ceased operation due to infrastructure aging
France	A10 pilot (Paris)	Freight & public transport	Wireless inductive	2	>100 (planned)	Evaluating decarbonization benefits for commercial fleets; next-gen coils	Planned for 2024; part of a larger national ERS plan

Sources [3]

hand, and simultaneously meet the climate and air quality targets, on the other hand [15]. Such a situation presents a special shortcut. Unlike the developed countries that have no choice but to retrofit their large, mature road systems, most Asian countries are currently building new roads and urban networks, thus allowing Electric Road Systems (ERS) to be integrated into the design process at a low cost, and potentially avoiding a generation of infrastructure that relies on fossil fuels. In this regard, a tailor-made and staged approach is needed to carry out the integration of ERS technology with the unique economic, geographic, and institutional environment of the region.

8.8.1 A Phased Regional Integration Model

An ER implementation strategy in emerging Asia should consider projects that bring the most economic and environmental benefits at the initial stage, allowing to entail the momentum and eliminating the threat of future investments.

- **Phase 1: High-Volume Economic Corridor Pilots**: The first one needs to be the electrification of short-to-medium-range, high-density freight routes that connect large industrial belts, ports and logistics centres. The given approach aligns with the national strategy of India to create priority Zero-Emission Truck (ZET) corridors, including the Delhi-Jaipur or Bengaluru-Chennai routes, through such initiatives as e-FAST (Electric Freight Accelerator) [15]. These pilots can be viewed as key demonstrations of the techno-economic functionality of ERS in the local environment and play a significant role in inviting the interest of the private sector. The hybrid technology model could be the best solution, using an in-ground system in the port access roads and the urban boundaries, and high-power catenary systems in the open intercity highways.
- **Phase 2: National Backbone Network Expansion**: Good results should be multiplied by connecting pilots, creating an interlinked network which forms a national or sub-regional backbone of electrified highways. This strategy is similar to the logic behind the UNESCAP Asian Highway Network; therefore, a system that promotes integration and facilitates the movement of freight over long distances across the key economic areas. This phase requires significant state capitalism and inter-jurisdictional management to implement.
- **Phase 3: Urban Public Transport Integration**: As the technology becomes more mature and the local experience grows, ERS can be implemented into the transport infrastructure of the major cities. New Bus Rapid Transit (BRT) projects should be the first targets since dedicated lanes will make it easier to install infrastructure and fully utilise the system. This step is based on the experiences of the freight corridors to improve the quality of air in the city and provide efficient and reliable mass transportation.

8.8.2 Institutional Roles and Collaboration

A successful ERS rollout depends on a clear and coordinated institutional framework that aligns the objectives of government, industry, and finance. Figure 8.5 illustrates a proposed collaborative structure.

The key institutional roles are defined as follows.

- **Ministry of Transport/National Highway Authority**: this body must lead the work, which is to determine and identify the priorities of the corridors, coordinate the multi-stage implementation, create technical principles that should be used to achieve interoperability, and organise the integration with the current road network.
- **Ministry of Energy/National Utilities**: This is the department that undertakes grid-impact studies to ascertain sufficient capacity, facilitates required grid upgrades, and devises appropriate tariff schemes and billing systems for electricity used through ERS.
- **Ministry of Industry and Finance**: The role of this ministry is to promote the supply and demand. It must encourage fleet conversion, through demand-side incentives- e.g. subsidies or tax reductions on buying ERS-capable vehicles, similar to India's PM -E-DRIVE scheme. Also, it has the mandate to develop the financial structure, such as public–private partnership (PPP) models and viability-gap funding, needed to appeal to the massive volume of private investment needed to develop infrastructure.
- **Regional Development Partners (e.g., ADB, UNESCAP)**: Multilateral institutions can play a crucial catalytic role by providing concessional financing for initial pilot projects, offering technical assistance for policy and standards development, and facilitating regional cooperation to ensure cross-border interoperability along international trade corridors [16].

8.9 Challenges and Mitigation Strategies

Although ERS has the potential to transform, the ubiquitous implementation of the system faces a lot of technical, economic, institutional, and social challenges. In spite of the fact that core technologies have reached a high level of maturity in pilot environments, the main obstacles to scaling are more of an economic and institutional nature. Such an active approach to finding and reducing these challenges is an essential part of the successful transition.

The key challenges and their corresponding mitigation strategies are summarised in Table 8.4.

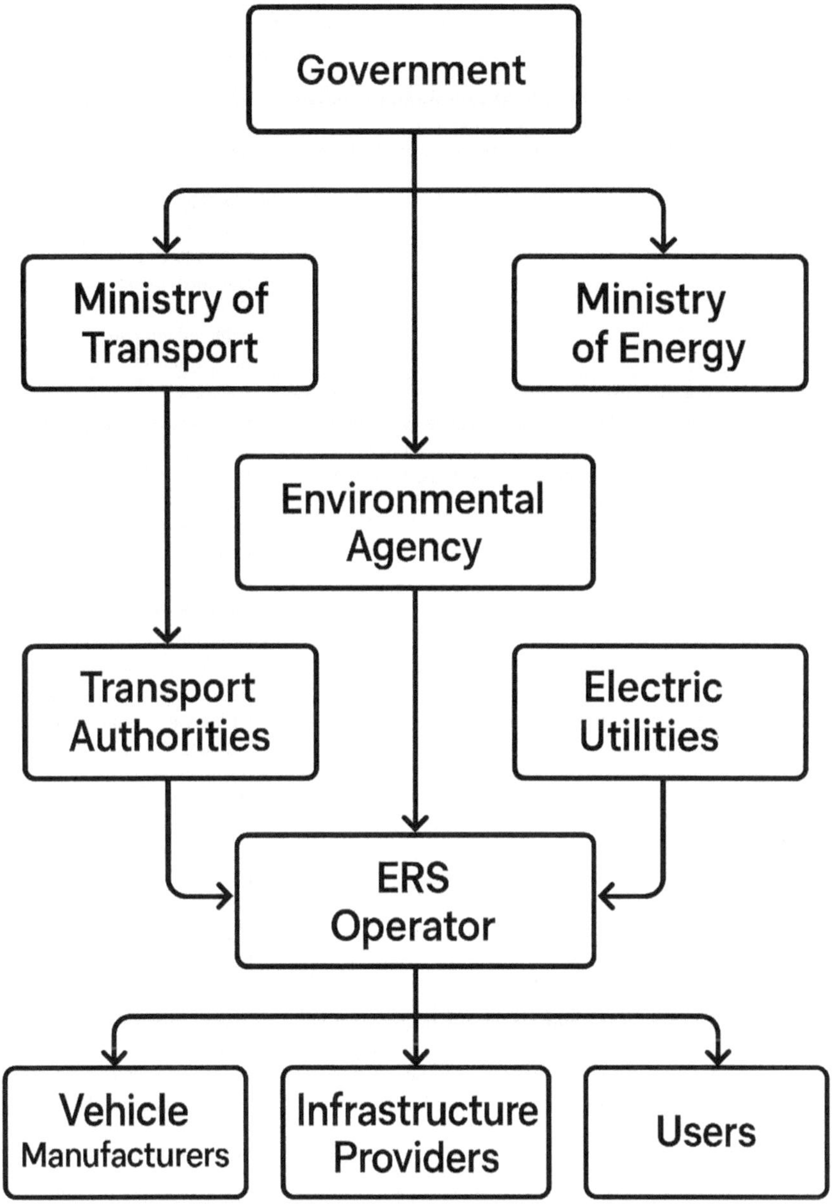

Fig. 8.5 Institutional framework for ERS integration in Asian transport systems

Table 8.4 Key challenges and proposed mitigation pathways for ERS deployment

Category	Specific challenge	Proposed mitigation strategy
Technical	**Lack of interoperability and standardization**: Systems from different vendors are incompatible, creating risk of vendor lock-in and fragmented networks	• Mandate open standards for physical interfaces, voltage, and communication protocols in public tenders • Establish national and international technical committees to drive standardization efforts • Promote modular designs for vehicle-side and infrastructure components
	Maintenance complexity: In-pavement systems (rail, inductive) can complicate road resurfacing and repair. Wear and tear on conductive components (pantograph carbons, rails) requires regular maintenance[3]	• Develop standardized, modular ERS components for easy replacement • Integrate ERS maintenance schedules with regular road maintenance programs • Utilize digital twins and predictive analytics for proactive maintenance planning
Economic	**High upfront capital expenditure (CapEx)**: The cost of electrifying roads is a major barrier for both public and private investors (€1.7–3.1 million/km)	• Develop Public–Private Partnership (PPP) models with clear revenue-sharing mechanisms (e.g., availability payments, user fees) • Provide government-backed viability gap funding or loan guarantees to de-risk private investment • Focus initial deployment on high-volume corridors to maximize revenue potential and economic return
	Uncertain revenue streams: Uncertainty about fleet adoption rates and willingness to pay makes it difficult to forecast revenue, deterring private finance	• Secure "anchor tenant" agreements with major fleet operators to guarantee baseline utilization • Implement government-led fleet procurement programs to create initial demand • Model dynamic pricing structures based on time-of-day and grid conditions

(continued)

Table 8.4 (continued)

Category	Specific challenge	Proposed mitigation strategy
Institutional	**Fragmented governance**: Responsibilities are split across multiple ministries (Transport, Energy, Finance, Environment) with potentially conflicting priorities, leading to policy delays	• Establish a high-level national ERS task force or dedicated authority with cross-ministerial representation to lead a coordinated strategy • Develop a single, unified national ERS roadmap with clear targets, timelines, and agency responsibilities
	Regulatory and legal gaps: Existing regulations for road construction, vehicle standards, and electricity sales may not accommodate ERS technology	• Proactively review and amend existing legislation to create a clear legal framework for ERS ownership, operation, and safety • Develop new regulations for ERS-specific issues like grid connection, billing, and data privacy
Social	**Fleet operator resistance**: Operators may be hesitant to invest in new, unproven vehicle technology due to perceived risks, higher upfront costs, and operational changes	• Provide strong demand-side incentives (subsidies, tax credits) to offset the higher initial cost of ERS-enabled vehicles • Conduct fleet-specific TCO analyses and demonstration projects to build operator confidence • Offer comprehensive driver training and technical support programs
	Public acceptance: Overhead catenary systems can face public opposition due to their visual impact on the landscape	• Conduct thorough public consultations and environmental impact assessments during the planning phase • Prioritize in-ground systems (rail or inductive) in scenic, urban, or residential areas where visual impact is a major concern

Sources [10]

8.9.1 Future Outlook

The development of the Electric Road Systems is sure to go beyond dynamic charging, incorporating a range of digital technologies to open new efficiencies and possibilities to transport operator and energy grid managers. The trend in the future is towards full electrification of the intelligent and interactive multimodal corridors.

8.9.2 Integration with Digital Twins and AI-Based Optimisation

The second direction in ERS management is the use of digital twins. A digital twin refers to a high-fidelity virtual representation of a physical system—in this case, the whole ERS network, including the road network, substation network and the fleet of vehicles that it interconnects [59]. This virtual model, continuously updated with real-time data from IoT sensors, enables powerful new applications.

- **Predictive Maintenance**: By simulating stress, wear, and energy flow, AI algorithms running on the digital twin can predict component failures before they occur, allowing for proactive maintenance that minimises downtime and improves system reliability.
- **Scenario Simulation**: Planners can use the digital twin to simulate the impact of network expansions, changes in traffic patterns, or extreme weather events, optimising design and operational strategies in a risk-free virtual environment.
- **AI-Based Fleet Optimisation**: Machine learning algorithms can leverage the holistic view provided by the digital twin to perform real-time, system-wide optimisation of fleet scheduling, vehicle routing, and energy consumption, far beyond the capabilities of current fleet management tools.

8.9.3 ERS and Vehicle-To-Grid (V2G) Logistics Nodes

Electric Road Systems (ERS) and Vehicle-to-Grid (V2G) technology, when combined, provide an intriguing prospect of how transportation energy management may be conducted in the future. Even though ERS helps to deploy smaller groups of vehicles, the cumulative storage capacity of a whole fleet of trucks or buses located at a logistics hub or depot is still a significant energy resource. By fitting the hubs with bi-directional chargers, the parked fleet can also be used to create a distributed battery and thus offer useful ancillary services to the electrical grid. This change will transform a logistics hub not only into an inactive energy consumer but also an active energy prosumer. The fleet will be able to recharge its batteries during low energy tariffs or when the amount of renewable generation is greater, and it can produce stored energy back to the grid during peak grid demand, thus helping to stabilise frequency and voltage and create a new revenue stream for the fleet operator. A digital twin framework is required here since, based on vehicle availability and state-of-charge, it predicts it and arranges a complex energy trading without a human touch and in an automated and optimised way.

8.9.4 Policy Synergy with Smart City Integration

The role of ERS technology as the building block of intelligent city infrastructure will become even more salient as it grows. Integrated ERS network offers data, connectivity, and energy flexibility, which are the prerequisites of the overall goals of sustainable urban mobility. The policy frameworks and digital platforms that are created to support ERS will have direct synergies with the initiatives that are discussed in Chap. 9, which explores further embedding of the transport systems into the smart city paradigm. ERS is not only a source of vehicle power, but it is also a way to a more connected, responsive, and resilient transportation ecosystem of the city.

8.9.5 Conclusion

Vision of the electric road systems combined with the transport and freight networks is a significant move towards the decarbonisation of high-utilisation transport areas. As has been shown in this chapter, ERS is not merely an alternative charging method, but it is a groundbreaking technology that is redefining the operational and economic paradigms of electric fleets. ERS is able to directly overcome significant issues with battery size, costs, weight, and charge downtime, which have historically slowed the electrification of buses and heavy-duty trucks due to the vehicle-centric paradigm of energy storage by shifting the paradigm to be infrastructure-centric in energy delivery.

Public transport applications have been analysed and found to have a strong synergy with Bus Rapid Transit systems, whereby ERS can develop electric trams on tyres that will combine both the efficiency of rail and the flexibility of buses. In the case of freight, the hybrid ERS trucks have allowed the so-called Spine and Ribs model, which provides an effective and scalable approach to electrifying arterial corridors, which form the backbone of modern logistics. Sweden, Germany, and South Korea case studies in the world, with varying technologies they have taken in the past, all coming to a similar conclusion: technology is viable, and the challenge is now on how to ensure the technology is here to stay economically and operationally.

In the case of the emerging world, like Asians, ERS offers an unprecedented frontier of skipping through the old-fashioned infrastructure and building new, clean transport systems entirely. However, the achievement of this potential, both in developing and developed economies, is dependent on a successful struggle with major non-technical barriers. Although technical barriers such as interoperability are essential, institutional and economic challenges, the costliness of high start-up capital, a disjointed governance framework, and the need to adopt novel regulatory and corporate paradigms are the most daunting obstacles to scale. Finally, the effectiveness of ERS will be determined by the presence of a strong political goodwill and the development of working platforms that bring together transport departments, energy producers, industry players, and financial institutions. By closing institutional gaps and exploiting the digital potential of ERS by integrating it with artificial intelligence, digital twins, and V2G systems, ERS can form a foundational infrastructural layer of sustainable, efficient, and networked transport networks of the future. This position as a facilitator of smart mobility offers a first-hand connection to the larger urban and smart city connections, discussed in the next chapter.

References

1. Wikipedia (2025) Electric road. https://en.wikipedia.org/wiki/Electric_road. Accessed 18 Oct 2025
2. International Energy Agency (IEA) (2025) Electric vehicles. https://www.iea.org/energy-system/transport/electric-vehicles. Accessed 18 Oct 2025
3. f3 Centre (2025) Electric road systems (ERS)—fact sheet. https://f3centre.se/en/fact-sheets/electric-road-systems-ers/. Accessed 18 Oct 2025
4. Delft University of Technology (2025) Electric bus charging infrastructures: technologies, standards, and configurations. https://research.tudelft.nl/files/210237345/Electric_Bus_Charging_Infrastructures_Technologies_Standards_and_Configurations.pdf. Accessed 18 Oct 2025
5. Siemens Mobility (2025) eHighway—dynamic charging infrastructure. https://www.mobility.siemens.com/global/en/portfolio/road/ehighway.html. Accessed 18 Oct 2025
6. Urban Transport Magazine (2025) Electric road for eBus charging on test in Lund, Sweden. https://www.urban-transport-magazine.com/en/electric-road-for-ebus-charging-on-test-in-lund-sweden/. Accessed 18 Oct 2025
7. f3 Centre (2025) Electric road systems—fact sheet (Misc-4_210510). https://f3centre.se/app/uploads/Electric-Road-Systems-fact-sheet_Misc-4_210510.pdf. Accessed 18 Oct 2025
8. DiVA Portal (2025) Socioeconomic analysis of electric road systems. https://www.diva-portal.org/smash/get/diva2:1658492/FULLTEXT01.pdf. Accessed 18 Oct 2025
9. Wikipedia (2025) Swedish transport administration electric road program. https://en.wikipedia.org/wiki/Swedish_Transport_Administration_electric_road_program. Accessed 18 Oct 2025
10. PIARC (2025) Electric road systems: a solution for the future. https://www.piarc.org/en/order-library/29690-en-Electric%20road%20systems:%20a%20solution%20for%20the%20future. Accessed 18 Oct 2025
11. Wikipedia (2025) Smart city. https://en.wikipedia.org/wiki/Smart_city. Accessed 18 Oct 2025
12. U.S. Department of Housing and Urban Development (2025) Transit-oriented development and affordable housing. https://archives.huduser.gov/PORTAL/pdredge/pdr-edge-featd-article-010725.html. Accessed 18 Oct 2025
13. Smart City Sweden (2025) EVolution road—the next generation electric roads. https://smartcitysweden.com/best-practice/412/evolution-road-the-next-generation-electric-road/. Accessed 18 Oct 2025
14. California Air Resources Board (2025) Sustainable heavy-duty initiatives for future technology. https://ww2.arb.ca.gov/our-work/programs/advanced-technology-demos-and-pilots. Accessed 18 Oct 2025

15. Asian Development Bank (ADB) (2025) ADB's work in sustainable transport. https://www.adb.org/what-we-do/topics/transport/overview. Accessed 18 Oct 2025
16. United Nations Framework Convention on Climate Change (UNFCCC) (2025) Implementation of the framework for capacity-building in developing countries. https://unfccc.int/sites/default/files/resource/sbi2024_02E.pdf. Accessed 18 Oct 2025
17. DiVA Portal (2025) Electric road systems. https://www.diva-portal.org/smash/get/diva2:1046753/FULLTEXT01.pdf. Accessed 18 Oct 2025
18. MDPI (2025) Assessing the socioeconomic impacts of an inductive electric road system. https://www.mdpi.com/2071-1050/17/5/2283. Accessed 18 Oct 2025
19. EVolution Road (2025) Electric road test site in Lund. https://www.evolutionroad.se/en/lund/. Accessed 18 Oct 2025
20. Lund University (2025) EVolution road—research project. https://portal.research.lu.se/en/projects/evolution-road. Accessed 18 Oct 2025
21. ResearchGate (2025) Feasibility of electric buses in public transport. https://www.researchgate.net/publication/323522526_Feasibility_of_Electric_Buses_in_Public_Transport. Accessed 18 Oct 2025
22. Trafikverket Bransch (2025) National roadmap for electric road systems. https://bransch.trafikverket.se/contentassets/becf6464a8a342708a143e7fe9e5f0ef/national_roadmap_for_electric_road_systems_20171129_eng.pdf. Accessed 18 Oct 2025
23. Transformative Mobility Initiative (2025) Catenary charging technology—IChargeHDV workshop. https://transformative-mobility.org/wp-content/uploads/2023/06/IChargeHDV_Workshop10_Catenary-Charging-Technology.pdf. Accessed 18 Oct 2025
24. Electreon (2025) Charge as you drive—France project. https://electreon.com/projects/france. Accessed 18 Oct 2025
25. Sustainable Bus (2025) Ground-charging system for the Tzen4 BRT line in Paris inaugurated. https://www.sustainable-bus.com/infrastructure/tzen-4-brt-paris-ground-charging-launch-alstom-kiepe-van-hool-ground-charging/. Accessed 18 Oct 2025
26. Nordic Energy Research (2025) Electric road systems. https://www.nordicenergy.org/wp-content/uploads/2019/10/Electric-roads_web.pdf. Accessed 18 Oct 2025
27. TransformCN (2025) eHighway—dynamic charging infrastructure overview. https://www.transformcn.com/uploads/20230908/10fd92fa7d9fd387a7f1dcb029d4b926.pdf. Accessed 18 Oct 2025
28. Integrated Transport Research Lab (ITRL) (2025) RENO–route-based ERS network optimization. https://www.itrl.kth.se/research/completed-projects/reno-route-based-ers-network-optimization-1.984073. Accessed 18 Oct 2025
29. IFEU (2025) Current technical findings on the eHighway system from field studies. https://www.ifeu.de/fileadmin/uploads/pdf/2022-10-24_ERS_Working_Paper_TechnicalAssessment_en_v2.pdf. Accessed 18 Oct 2025
30. Centre for Sustainable Road Freight (2025) eHighway virtual tour—ELISA project. https://www.csrf.ac.uk/wp-content/uploads/2020/11/5.-Virtual-Tour-eHighway-ELISA_SMO_GS_MS.pdf. Accessed 18 Oct 2025
31. Siemens Press (2025) Third eHighway field trial starts in Germany. https://press.siemens.com/global/en/pressrelease/test-field-ehighway-project-ewaybw. Accessed 18 Oct 2025
32. Smartroad Gotland (2025) Home Page. https://www.smartroadgotland.com/. Accessed 18 Oct 2025
33. Smart City Sweden (2025) Wireless electric road charges vehicles as they drive. https://smartcitysweden.com/best-practice/409/wireless-electric-road-charges-vehicles-as-they-drive/. Accessed 18 Oct 2025
34. Electreon (2025) Findings from the world's first public wireless electric road for heavy-duty commercial vehicles. https://electreon.com/articles/worlds-first-public-wireless-electric-road. Accessed 18 Oct 2025
35. Cision (2025) Installation of the world's first wireless electric road for trucks and buses begins—Visby, Sweden. https://mb.cision.com/Main/18921/2959916/1140125.pdf. Accessed 18 Oct 2025

36. Fleetio (2025) Fleet management software for smarter operations. https://www.fleetio.com/solutions/fleet-management-software. Accessed 18 Oct 2025
37. Geotab (2025) One platform—total fleet management. https://www.geotab.com/. Accessed 18 Oct 2025
38. IFEU (2025) Opportunities for electric road systems in road freight economics. https://www.ifeu.de/fileadmin/uploads/Publikationen/Mobilit%C3%A4t/CollERS-2-Discussion-paper-3-Opportunities-for-electric-road-systems-in-road-freight-economics.pdf. Accessed 18 Oct 2025
39. PubMed Central (PMC) (2025) Cost, energy, and carbon footprint benefits of second-life electric vehicle battery use. https://pmc.ncbi.nlm.nih.gov/articles/PMC10339184/. Accessed 18 Oct 2025
40. Eva-Last USA (2025) The role of smart cities in future urban planning. https://www.eva-last.com/us/the-role-of-smart-cities-in-future-urban-planning/. Accessed 18 Oct 2025
41. KPF Urban Interface (2025) The smart(er) city. https://ui.kpf.com/smarter-city. Accessed 18 Oct 2025
42. ResearchGate (2025) Urban planning and smart cities: interrelations and reciprocities. https://www.researchgate.net/publication/230851445_Urban_Planning_and_Smart_Cities_Interrelations_and_Reciprocities. Accessed 18 Oct 2025
43. Metropolitan Atlanta Rapid Transit Authority (MARTA) (2025) Transit-oriented development guidelines. https://itsmarta.com/uploadedFiles/More/Transit_Oriented_Development/TOD%20Guidelines%202010-11.pdf. Accessed 18 Oct 2025
44. ResearchGate (2025) KAIST wireless electric vehicles—OLEV. https://www.researchgate.net/publication/375560361_KAIST_Wireless_Electric_Vehicles_-_OLEV. Accessed 18 Oct 2025
45. MDPI (2025) Adopting an actor analysis framework to a complex technology innovation project: a case study of an electric road system. https://www.mdpi.com/2071-1050/12/1/313. Accessed 18 Oct 2025
46. California Energy Commission (2025) Zero-emission medium- and heavy-duty infrastructure loan pilot project. https://www.energy.ca.gov/programs-and-topics/programs/clean-transportation-program/zero-emission-medium-and-heavy-duty. Accessed 18 Oct 2025
47. NITI Aayog (2025) Transforming trucking in India. https://www.niti.gov.in/sites/default/files/2023-02/ZETReport09092022.pdf. Accessed 18 Oct 2025
48. South Asia Subregional Economic Cooperation (SASEC) (2025) Historic opportunity to link South and Southeast Asia: ADB brief. https://www.sasec.asia/index.php?page=news&nid=305&url=historic-opportunity-to-link-sa-sea-adb. Accessed 18 Oct 2025
49. ISEAS–Yusof Ishak Institute (2025) Promoting cross-border connectivity in Asia: the role of the Asian development bank. https://www.iseas.edu.sg/wp-content/uploads/2022/08/ISEAS_EWP_2022-4_Menon.pdf. Accessed 18 Oct 2025
50. Principal Scientific Adviser, Government of India (2025) India's priority corridors for zero emission trucking. https://psa.gov.in/CMS/web/sites/default/files/publication/India%27s%20Priority%20Corridors%20for%20Zero%20Emission%20Trucking_1.pdf. Accessed 18 Oct 2025
51. Times of India (2025) Government identifies 10 highway segments for zero-emission truck movement. https://timesofindia.indiatimes.com/india/government-identifies-10-highway-segments-for-zero-emission-truck-movement-the-move-to-help-india-reach-its-climate-goals/articleshow/121259365.cms. Accessed 18 Oct 2025
52. Climate Fact Checks (2025) India charts course for greener freight: ten highway corridors designated for zero-emission trucks. https://climatefactchecks.org/india-charts-course-for-greener-freight-ten-highway-corridors-designated-for-zero-emission-trucks/. Accessed 18 Oct 2025
53. Nokia (2025) How digital twins are driving the future of engineering. https://www.nokia.com/thought-leadership/articles/how-digital-twins-driving-future-of-engineering/. Accessed 18 Oct 2025
54. Automate Show (2025) The future of business—your guide to digital twin technology. https://www.automateshow.com/blog/your-guide-to-digital-twin-technology. Accessed 18 Oct 2025

Chapter 9
Urban and Smart City Integration of Electric Road Systems (ERS)

9.1 Introduction

Electric Road Systems (ERS) can be described as a paradigm shift in the electrification of transportation, whereby it does not follow the conventional model of static charging, but allows mobile, in-motion provision of energy. The urban application of ERS is not a new technology but a planned infrastructure that is meant to serve a large number of urban vehicles in total, which includes a fleet of public transportation, electric buses, as well as services needed by the city, such as garbage trucks and delivery vans, and finally, it is a fleet of individual passenger vehicles. The system works by transferring electrical energy using the road infrastructure to a moving vehicle using one of three main archetypes of technology: conductive overhead lines (catenaries), conductive in-road rails, or wireless inductive coils under the pavement. More importantly, ERS is not conceived as an extended network that covers all the streets in a city, but as a limited range extender that will be installed strategically in the major streets, logistical and transport pathways within a city. This arrangement allows the vehicles to charge as they pass through these main routes and uses smaller onboard batteries to make use of secondary routes and the last part of their route [1].

A combination of immediate difficulties that characterise modern urban areas determines the necessity of implementing such a transformative framework into the city environment. First, air pollution that is caused by long-term traffic jams is known to be a major health and economic productivity risk to the masses. In densely populated cities, the amount of pollutants (carbon monoxide (CO), nitrogen oxides (NOx), and particulate matter (PM 2.5)) produced by vehicles may be more than twice as much as found in traffic with a low level of congestion, thus being a direct cause of respiratory diseases and early death. This would allow the mass electrification of high-mileage, high-utilisation urban fleets, which are sources of this pollution disproportionately, and would directly address a major source of environmental degradation and reflect the work done globally on the problem of air quality in cities. According to the United Nations Environment Programme (UNEP), one of the most important

R. Agnihotri, *Electrified Road Systems (ERS) for Emerging Asia*, Springer Tracts on Transportation and Traffic 24, https://doi.org/10.1007/978-3-032-16927-3_9

aspects of making cities more sustainable is the reduction of emissions caused by road transport [2].

Second, a major challenge to urban energy grids is the uncoordinated and massive electric vehicles (EVs) on a large scale. The traditional depot-based charging system, especially with large fleets of electric buses, may create super-peaks of electricity consumption. Empirical research and data show the possible impact of electrifying a bus depot by 17 or even 20 or 30% on the peak load of a local substation, and that evening peak periods may increase by 20 or 30%, triggering grid instability and leading to expensive infrastructure upgrades. The key point is that ERS changes this balance by spreading the charging load not only geographically along the road system of the city, but also in time, all throughout the day. This change causes EV charging to become a disruptive event that causes peak loads, but becomes a manageable, distributed and potentially flexible load which can be intelligently coordinated with the grid [3].

Third, ERS reduces the internal constraints of battery dependency to the EV transition. The battery is the only component of an EV that is very costly and the heaviest part, and its manufacturing depends on a chain system of essential minerals that are full of geopolitical and environmental hazards. In high-utilisation city cars, the need to travel long distances requires large, heavy, and expensive batteries, which increases the capital cost of a vehicle, impairs energy efficiency, and, in commercial cars, affects payload. ERS breaks this paradigm by separating the range of a vehicle in operation from the size of the onboard battery. The system can allow the use of much smaller batteries and consequently lower the cost, weight, and reliance on precious resources of vehicles by offering constant in-motion energy and therefore allowing the deployment of vehicles that are much smaller. This systemic change is a profound modification of the economic and logistical calculus of urban electrifying its fleets, which makes it more feasible and sustainable.

ERS, as a component of the larger concept of smart mobility and sustainable urban development, also represents a foundational technology enabling a systemic shift to a cleaner, more efficient and highly integrated urban transport. Its practice is in line with the targets of the international systems like the UN Sustainable Development Goal 11 (SDG 11), which is currently in support of free, accessible, and affordable transport systems for everyone. The approach to e-mobility adopted by institutions, including the United Nations Economic Commission for Europe (UNECE), is systems-based and cross-sectoral and describes ERS as a pivotal infrastructural nexus, differentiating between the physical and digital places of convergence between the energy and transport sectors.

The chapter shows the integration of ERS in the city ecosystem and smart city infrastructures. It starts off by placing ERS in the context of a multi-layered system of a smart city, outlining its purpose as a cross-cutting technology. It further studies the technological foundation of ERS, looking at traffic and energy management systems based on IoT and essential data and communication structures to guarantee interoperability. Then, the chapter examines how ERS can be practically integrated with urban system mobility and the necessary urban planning and infrastructure preparedness. It also examines the policy and governance processes that are necessary for the successful deployment. The research is based on a survey of the global case studies, on which the conceptual framework of the ERS-enabled smart cities, especially in the Asian environment, is offered. The chapter ends by evaluating the possible challenges and opportunities of the integration of urban ERS, thus preconditioning the further consideration of the new technologies in the field.

9.2 ERS in the Smart City Ecosystem

In order to imagine the capacity of Electric Road Systems to transform an urban setting, it is necessary to imagine the smart city as a system of layers, first. One of the most commonly used models describes a smart city in terms of four major layers: a Mobility layer, which controls movement of people and goods; an Energy layer, which manages the production, distribution, and consumption of power; a Governance layer, which deals with policy, regulation and government; and an integrating Data layer that cuts across and connects all these other three layers. These layers are not independent silos, but they are highly interdependent, and the data and connectivity act as the connective tissue, and this is where intelligent, cross-domain optimisation occurs. ERS stands alone in not being a layer-specific component but as an underlying technology, which cuts across and also actively works across all four layers, serving as a physical-digital backbone to the urban energy-responsive mobility nexus.

In the Mobility layer, the ERS is not involved in the provision of energy alone. It essentially increases the efficiency of the urban transport systems by allowing them to charge on the fly. In the case of electric public transport, ERS allows the continuous operation of a vehicle on a high-frequency route without having to interrupt its operation during stops at depots/terminals, which increases vehicle availability and service quality, and subsequently, directly leads to a more efficient urban logistics system. Moreover, the information obtained through the ERS network can be integrated into the Intelligent Transportation System (ITS) of a city to provide real-time traffic control, give priority to electric vehicles in special lanes, and optimal path to facilitate the overall reduction of congestion [4].

The ERS acts as an intelligent interface between the transport sector and the smart grid in the Energy layer, which is a critical interface. This is in contrast to the unscheduled and usually disruptive burden posed by traditional EV charging, which is a distributed, flexible, and controllable energy asset [5]. It is also able to contribute actively to grid balancing, unlike the severe peaks on local substations that could be caused by depot charging. In times of excess renewable power (e.g. when the sun is blazing in daytime), the ERS can absorb this clean energy by encouraging charging. In contrast, the grid stress may be modulated during times of power demand. Grinding Vehicles with ERS communication capabilities could one day become a distributed grid of moving batteries that deliver ancillary services, helping to make the grid stable and resilient through the application of vehicle-to-grid (V2G) communication standards like ISO 15118.

Implementation and usage of ERS requires new structures in the Governance layer. The expensive capital and nature of a public good make the infrastructure an excellent candidate for innovative public–private partnership (PPP) models, whereby the capital and the operational expertise of the industry are employed under the control of public regulatory authorities [6]. ERS would have needed a digital governance framework to control access, billing and performance. This layer is effective in terms of connecting the physical infrastructure with the financial, legal, and administrative systems of a given city, introducing digital permits of authorised vehicles, implementing dynamic pricing schemes, and providing a transparent policy of data ownership and privacy.

The Data layer represents the brain of the ERS-enabled smart city. The system is also data-intensive, based on a continuous real-time stream of data supplied by a large network of sensors on the roadway, roadside devices, and vehicles themselves as members of the Internet of Things (IoT) [7]. This is not only needed in the fundamental working functions of the ERS, like the dynamic activation of power segments, but also in the higher-level functions of the ERS in traffic management, energy monitoring, predictive maintenance and urban planning.

A smart city can be described by the ability to optimise the use of resources through connectivity via data, and transport and energy are the largest systems of resources in the city, traditionally operated separately but now linked directly, albeit uncontrolled, by the use of electric vehicles. ERS provides the mandatory intelligent and controlled interface. The mobility layer (the road) is physically connected to the infrastructure but is digitally controllable as part of the energy layer (the smart grid) [1]. By acting on information about the ERS (e.g., vehicle presence, energy demand) and the grid (e.g., price signals, renewable energy availability), the system will be used to make real-time coordinated decisions to bring the whole connected fleet of vehicles under control by the transport authority and the energy utility in a city (Fig. 9.1).

9.3 IoT-Based Traffic and Energy Management

The functional basis of an urban Electric Road System is a complex cyber-physical system coordinated by an extensive network of Internet of Things (IoT) gadgets that allow regulating the transport traffic and energy delivery in real time. This system includes sensors and edge devices that sense the environment, communication protocols that transmit the data, and smart algorithms that make decisions.

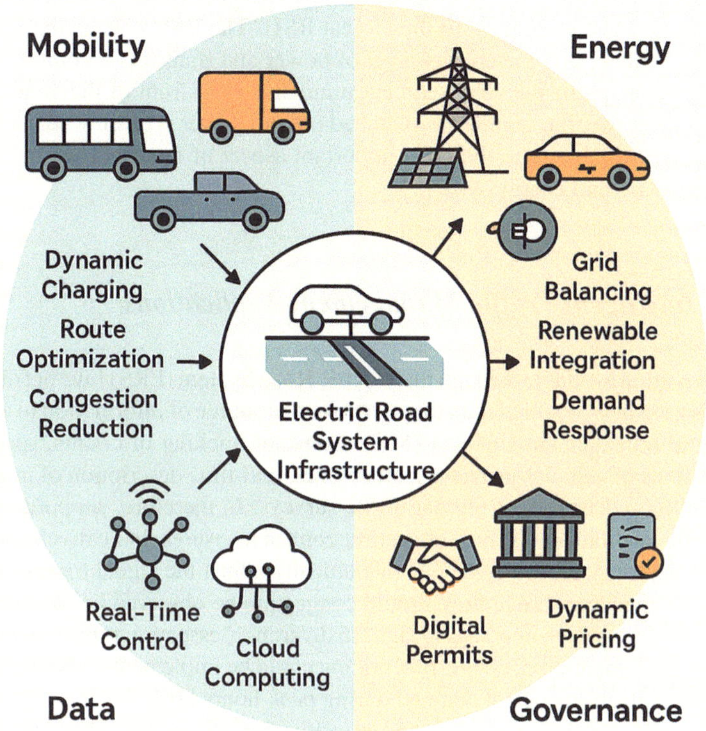

Fig. 9.1 ERS integration across smart city layers

9.3.1 *The Cyber-Physical System of Urban ERS*

The ERS is based around the sensors and edge devices that form part of the perception and local control layer. The sensors are deployed all over the infrastructure: in-road magnetic or optical sensors point to the exact position, speed and type of an approaching vehicle, power-monitoring sensors within the ERS power electronics monitor the energy transfer in real-time, and on-board sensors in the car (including GPS to determine location and battery management systems to monitor state-of-charge) provide important data on the vehicle side. RSUs are also needed to support low-latency operations, such as activating a power segment nearly instantly when a vehicle travels across it, or simultaneously detecting a fault condition that would necessitate segment closure to ensure safety [8].

This local processing is motivated by a sustained vehicle-to-infrastructure (V2I) channel of communication. When an ERS-equipped vehicle comes to an electrified lane, the on-board unit (OBU) wirelessly transfers its unique identifier, authentication credentials and energy request to the closest RSU. The RSU checks the authorisation of the vehicle, sets the correct level of power and transmits a command to the activation of the particular road segment immediately in front of the vehicle. With the car on the move, the segment is switched off and power is given when and where it is required, which is also the most important aspect of modern ERS technology, both safety-wise and energy-wise [9].

9.3.2 Intelligent Traffic Management Applications

The information available through the Electric Road System (ERS) Internet of Things network is a heavy stream of data that is a powerful source of information to a municipal Traffic Management Centre (TMC). Constant tracking of counts, speeds and classifications of vehicles provides a much finer, real-time description of traffic than traditional loop detectors or regular traffic surveys. It, therefore, supports sophisticated traffic load-balancing and congestion-control measures. Indicatively, in a more advanced deployment, the TMC can dynamically adjust the signal timing at downstream crossings to improve flow should congestion be observed in one lane of the ERS, or can use variable messaging signs to divert non-essential traffic [4]. In a more advanced implementation, congestion pricing could be implemented dynamically on the ERS lane itself to control demand during peak hours [10].

In addition, predictive routing and prioritisation of electric vehicles (EVs) is possible because of the presence of historical and real-time data of ERS use. With such data fed into artificial-intelligence and machine-learning models, the TMC will be able to predict the traffic patterns and areas of congestion. This allows providing optimised routes to ERS-equipped vehicles, especially commercial ones, to the maximum on the electrified corridors, which minimises battery usage and improves operational efficiency [7]. Most importantly, the system also allows active prioritisation of particular vehicle classes. In times of maximum traffic jams, it is possible to program signal systems to form a green wave of electric buses or emergency vehicles using the ERS lane, which will guarantee the quality of the work of public transport and increase the time of emergency rescue.

9.3.3 Smart Grid Integration and Energy Synchronisation

The ERS, combined with the city's smart grid, is indispensable due to the maintenance of grid stability and the fullest use of renewable energy. The ERS control system should be able to communicate with the Supervisory Control and Data Acquisition (SCADA) platform and local substation monitoring instruments of the utility, hence providing the utility with real-time insight into the ERS network as a predictable and manageable load. Instead of responding to a sudden increase in demand, the grid operator may predict the energy demand of the ERS with the help of the traffic data and deal with its influence in advance [5].

This controllability allows the ERS to be actively involved in the demand-response programmes. Upon stress of the grid, the utility may send a signal to the ERS control centre to temporarily reduce the maximum power output of charging lanes, which would shed load systematically. On the other hand, and possibly more significantly, the ERS is able to take in excess renewable generation. When there is a surplus of solar or wind production during the time when the demand is low relative to the supply, the ERS may be asked to make lower charging rates. This incentive stimulates ERS-equipped vehicles to use electrified corridors, which in effect makes the aggregate batteries in the fleet a distributed energy sponge, which absorbs extra clean power and does not get curtailed, greatly improving the economics of renewable generation [11].

The ERS IoT architecture creates a high-resolution real-time digital twin of traffic dynamics and energy consumption on major arteries of a city, where conventional urban management has normally been reactive based on sparse and delayed data [12]. By contrast, an ERS-equipped corridor records data about all the vehicles connected to it: what they are, how fast they are, and how much energy they are actually consuming, on a second-by-second basis [9]. When summed together, such a stream of data provides a fine-scale dynamic description of the performance in a corridor. Algorithms using artificial intelligence to analyse this vast data can go beyond mere monitoring to make predictions about future conditions, including an impending congestion or an upcoming surge in power demand [7]. This predictive power will enable city operators in the traffic and energy sectors to take proactive measures instead of only being reactive to changes in the environment (Table 9.1).

9.4 Data Architecture and Communication Framework

A powerful, safe and interoperable data architecture is the key to the successful implementation of an Electric Road System into a smart city. This framework will have to coordinate the flow of information between vehicles, roadside infrastructure, and central city management platforms, and thus, the gap between the different worlds of automotive engineering, energy management, and information technology will be narrowed. This architecture can best be thought of as a multi-tiered system that is supported by a set of essential international standards that ensure functionality and avoid vendor lock-in.

Table 9.1 Smart components of ERS-enabled city infrastructure

Component	Type	Function	Integration role
Vehicle detection sensors	In-road (magnetic/optical)	Detects presence, speed, and class of vehicle approaching an ERS segment	Triggers V2I communication for dynamic lane activation and provides raw traffic data
Roadside unit (RSU)	Edge computing device	Processes V2I communication, manages local ERS segment activation, and authenticates vehicles	Enables low-latency control, reducing reliance on the central cloud for critical functions
On-board unit (OBU)	In-vehicle device	Transmits vehicle ID, battery status, and energy request to the RSU	Serves as the vehicle-side component of the V2I communication loop
Smart meters (in ERS)	Energy monitoring hardware	Measures power transfer to each vehicle for billing and grid monitoring purposes	Provides granular data for energy management, demand forecasting, and transaction logging
ERS control centre	Cloud/central server	Aggregates data from all RSUs, manages billing, and interfaces with the TMC and utility	Acts as the central nervous system for city-wide ERS operations and data exchange
Traffic management centre (TMC)	City infrastructure	Ingests ERS traffic data to optimize signal timing, manage congestion, and provide driver information	Links ERS operational data to the city's overall traffic strategy and control systems
Utility control centre (SCADA)	Energy grid infrastructure	Monitors the ERS load on substations and sends price or control signals for demand response	Integrates the ERS into the smart grid, enabling it to function as a manageable asset

9.4.1 Structure of the Integrated Data Network

An ERS of a city is modelled by the data architecture of a typical four-layered Internet of Things framework, which guarantees a logical division of roles and efficient data processing:

1. **Perception Layer**: This forms a physical-digital interface where data is generated. It consists of On-Board Units (OBUs) in cars, collecting and transferring vehicle status information, and a myriad of sensors built into the road infrastructure, including vehicle detectors, power meters, and environmental sensors, that hand over real-time operational data [13].

2. **Network/Communication Layer**: The layer is in charge of conveying data between the perception layer and the processing layers, and it uses a non-homogeneous combination of wireless technologies, such as Dedicated Short-Range Communications (DSRC) or a cellular analogue, Cellular Vehicle-to-Everything (C-V2X), to support low-latency communication between vehicles and infrastructure. In cases where communication needs to be made between the roadside infrastructure and the central cloud, high-bandwidth solutions like 5G or fibre optics are used [13].
3. **Edge/Fog Layer**: This processing layer is an intermediate layer that is comprised of Roadside Units (RSUs) and other local controllers. These devices can process data with time-critical requirements locally, i.e. vehicle authentication and immediate activation/deactivation of ERS segments, which lowers latency and offloads the main cloud, improving system responsiveness and reliability [8].
4. **Cloud/Application Layer**: On top of the architecture, there are central cloud servers. This layer integrates data within the whole ERS network to do analytics across the city, process billing and payments, long-term planning, and system optimisation. It contains the applications and dashboards used by operators and fleet managers of the city, as well as by the citizens, and acts as the interface to other smart city platforms [14].

9.4.2 Key Communication Standards and Protocols

The critical innovation represented by a smart urban Electric Road System (ERS) is not seen in the specific hardware constituents but in the creation of a protocol bridge that would allow well-established standards within three different industrial sectors, automotive, electric-vehicle charging, and power-grid infrastructure to inter-languages with unimpeded faithfulness. This interoperability forms the key to the realisation of total management that is intelligent throughout the system.

- **ISO 15118 (Vehicle-to-Grid Communication Interface)**: This global standard forms the basis of safe and comprehensible communication between electric vehicles (EVs) and the charging system. In the ERS setting, it regulates the digital handshake between an On-Board Unit (OBU) and a Roadside Unit (RSU) of a vehicle. Its salient contributions include the support of the so-called Plug '&' Charge feature, thus making the process of authentication and authorisation automatic and not requiring driver intervention, and creating a secure and encrypted channel with the help of a Public Key Infrastructure (PKI). Notably, the most recent ISO15118-20 offers the required message sets needed to achieve a two-way power transfer (V2G), thus enabling the use of ERS-equipped vehicles as distributed energy sources that can be used to perform grid operations [15].
- **OCPP 2.0.1 (Open Charge Point Protocol)**: OCPP 2.0.1, though originally designed to control a stationary charging station, is also necessary to control the ERS infrastructure itself. It provides a standardised lexicon of communication between single ERS power-supply units, which are called charge points,

and the central management system (CSMS), which is run by municipal authorities or third-party operators. The protocol delivers adequate services including remote monitoring, diagnostics, firmware updates, billing, logging transactions, and dispatching of advanced smart-charging commands (e.g. reduce power out to 80%). The adherence to OCPP 2.0.1 is essential to allow a city to buy ERS hardware supplied by different vendors and operate it via a single and standard software platform, which will reduce the risk of vendor lock-in [16].

- **IEC 61850 (Communication Networks and Systems for Power Utility Automation)**: This standard is the international standard in grid-level communication, especially in substations and to manage Distributed Energy Resources (DERs). To be a true smart-grid asset, the central control system of an ERS has to connect to the utility control centre, using IEC 61850. This functionality would enable the ERS to be modelled as a logical node in the digital twin of the grid, and provide the high-speed, reliable, and secure communication of important grid functions like load-shedding coordination in the case of an emergency or precise modulation of the ERS power consumption to stabilise grid frequency.

9.4.3 Role of Advanced Digital Technologies

Beyond these core standards, several other technologies are vital for the data architecture:

- **Cloud Computing and AI Analytics**: City-scaled ERS network requires data scalability that is found in cloud computing and processing. This data is analysed by artificial intelligence and machine-learning programs running in the cloud over a spectrum of applications, including predicting traffic congestion and energy demand and optimising machine maintenance directly by detecting predictors of machine failure [7].
- **Blockchain for Transaction Transparency**: The blockchain technology can provide an opportunity in terms of transactions that are involved in energy exchange, and is linked with high volumes of micro-transactions. Blockchain improves trust and security in the billing processes and generates a verifiable audit trail of carbon-credit schemes or peer-to-peer energy exchanges between vehicles by offering a decentralised, irrevocable, transparent ledger that documents every individual energy transfer event [17].
- **Cybersecurity Framework**: The interconnection that characterises the ERS makes it a high-value target of cyberattacks. There must be a developed cybersecurity framework, including that identified by the National Institute of Standards and Technology (NIST). This framework will have a multilayered, defence-in-depth approach, which involves end-to-end encryption (required by ISO 15118), strict authentication of all devices and users, constant network monitoring to identify intrusions, and frequent security audits and penetration testing (Fig. 9.2).

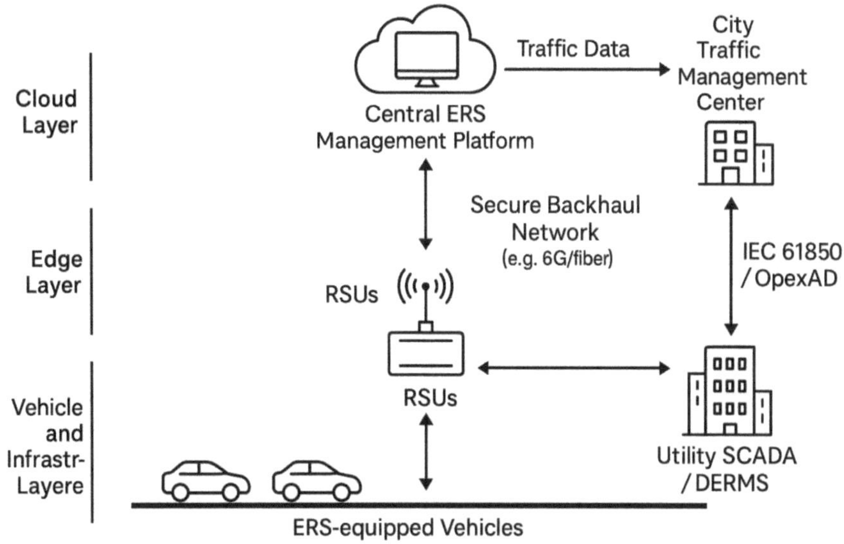

Fig. 9.2 Data flow diagram for ERS—city integrated network

9.5 Integration with City-Level Mobility Systems

An Electric Road System can only be said to be of substantive value once fully incorporated into a heterogeneous mobility ecosystem within a city as a backbone that is both more efficient, economically viable and increases the quality of service provision across various forms of transport simultaneously. The ERS is to be not only a siloed project, but it should be modelled as a strategic enabler of electrification of public transport, optimisation of municipal and commercial fleets, and strengthening of intermodal connectivity.

9.5.1 *Electrifying Public Transport Corridors*

The primary and most immediate use of urban ERS is the electrification of public bus routes. Buses that are operated with high frequency are ideal candidates, since they deal with heavy buses that move along fixed routes and cover great mileage in a day. Transit agencies can fundamentally restructure their electric bus fleets by installing ERS segments, either inductive or conductive, along these core corridors, especially on the parts of the line where energy demand is high (steep slope or long, uninterrupted stretches). The model allows achieving a variety of advantages: The cost of acquiring vehicles at the start is reduced, since the vehicles can operate continuously without large batteries with significant weight, resulting in a better quality and appeal of the transport in general; the frequency of serving people is increased, and they can use the same number of vehicles to provide a higher number of services. The On-Line Electric Vehicle (OLEV) system, which is currently implemented in bus fleets of the South Korean cities of Busan and Daejeon, can be regarded as an innovative application of the concept into reality, providing proof that dynamic wireless charging on urban transit is a practical idea [18].

9.5.2 *Optimising Municipal and Commercial Fleets*

In addition to the traditional public transport, the Electric Road System (ERS) offers an attractive platform for the electrification of high-utilisation urban fleets. Examples of municipal services, such as garbage collection and more general commercial operations, including package delivery and ride-hail, feature vehicles travelling over long distances in a day on predictable or semi-predictable routes where the physical mass of a conventional large battery significantly reduces the legal payload of the truck as well as its operational efficiency. The adoption of ERS can also be used to equip smaller, lighter batteries that restore payload capacity and increase economic viability. In the case of a delivery van and taxis, the ERS would eliminate the so-called refuelling downtime, which cuts off revenue-generating hours, making the electrification of fleets more viable in general.

9.5.3 *Supporting Micro-mobility and Intermodal Connectivity*

A strategically planned ERS network also plays a crucial role in strengthening a city's intermodal transport system and solving the critical "first and last mile" problem. ERS corridors should be designed to connect residential and commercial zones with major intermodal hubs, such as metro stations, railway terminals, and central bus interchanges. This creates seamless, electrified pathways for feeder services that bring people to and from high-capacity public transport.

A literature or well-planned ERS network plays a major role in the strengthening of an intermodal transport system of a city and fulfils the longstanding issue of the last and first mile. ERS corridors should be designed to connect residential and business areas with key intermodal terminals such as metro stations, railway stations, and central bus exchanges, thus creating smooth and electrified corridors of feeder routes that carry passengers to and away high-capacity mass transit. Whereas ERS technology does not involve direct charging of small personal mobility devices like e-scooters or e-bikes, the large power infrastructure required to implement an ERS corridor, including substations, transformers and high-capacity cabling, can be used to provide ancillary mobility modes. This infrastructure can be used to develop combined charging and docking stations of shared micro-mobility fleets that are located in strategic locations along the ERS corridor, especially near transit stations. This would create a synergistic association through which the macro-infrastructure of the ERS supports the micro-mobility networks and thus provides the much-needed first- and last-mile connectivity and allows urban residents to travel on a fully electrified, multi-mode journey [19].

Vehicle availability and the cost of operation are inherently restrictive to the operational efficiency of any urban mobility service. In the case of electric vehicles (EVs), the constraints involved will be the length of charge, and operational costs will be determined by the cost of electricity and battery lifecycle costs. Having a single ERS corridor serving buses, delivery vans, and taxis simultaneously will improve the operational and economic performance of public transportation, urban logistics, and ride-hailing operations. This creates a multiplier effect, or a mobility service multiplier, such that one infrastructure investment could be spread to the entire urban mobility ecosystem with a series of positive effects (Table 9.2).

Table 9.2 Urban mobility use-cases integrated with ERS

Use-case	Vehicle type	ERS integration strategy	Key benefits
Public transit	City buses	Inductive or catenary segments on high-frequency routes and at bus stops	Smaller batteries, reduced vehicle weight, continuous operation (no charging downtime), lower capital cost per bus
Municipal services	Garbage trucks	Conductive rail or inductive ERS on arterial roads leading to waste processing facilities	Increased payload (due to smaller battery), predictable charging on fixed routes, reduced operational costs and emissions
Urban logistics	Delivery vans	ERS on major commercial corridors connecting distribution centers to the city core	Extended daily range, reduced need for mid-day charging, enables 24/7 delivery cycles with smaller fleets
Ride-hailing/taxis	Passenger cars	ERS on high-traffic roads, airport access routes, and central business district loops	Reduced driver downtime for charging, increased daily revenue-generating hours, lower vehicle acquisition/lease cost
Intermodal hubs	All vehicle types	ERS corridors terminate at metro/train stations; leverage ERS power infrastructure for micro-mobility charging docks	Enables seamless multi-modal journeys, solves the first/last mile problem with a fully electrified transport chain

9.6 Urban Planning and Infrastructure Readiness

The adoption of Electric Road Systems goes beyond the technological upscale; it is an incidental phenomenon, an act of urban change, which requires active, cross-system thinking. ERS cannot be a retrofit, but should be built into the very fabric of a city's long-term development plan, and it has to impact the master plans, zoning codes, and the coordination of the municipal agencies. This kind of undertaking needs to take a holistic approach that simultaneously considers the physical, institutional, and environmental aspects of infrastructure preparedness.

9.6.1 Incorporating ERS in Urban Master Plans

The implementation of ERS must be guided by the strategic vision of a city in terms of growth and movement, to ensure that the system has a desired efficacy. This starts with its formal incorporation in the city's master plans. City-wide analyses of the salient corridors should be conducted by the planners to mark them as priority candidates for ERS electrification. These corridors usually cover large arterial roads connecting residential areas with working centres, high-volume freight routes connecting ports or logistics centres with the city and the most heavily used routes of the general transportation. A city expresses a long-term commitment, thus providing certainty to the individual investment and harmonising the future infrastructure spending between the departments.

The strategic designation should be supplemented with the relevant changes in zoning structures and land-use regulations. The current zoning codes usually do not include dynamic charging infrastructure classifications. Municipalities need to develop new ordinances that clearly state ERS is an authorised activity in appropriate zoning areas, which would help in the protracted procedure of obtaining a permit [20]. In addition, zoning may serve as a proactive tool to drive complementary development. An example is that municipalities might offer density bonuses or expedited permission to the construction of logistics or intermodal transit centres or mixed-use residential developments located on an E-Corridor, and connected to ERS infrastructure.

9.6.2 Institutional Coordination and Physical Modifications

Implementation of ERS is a multi-disciplinary activity that cuts across traditional silos of municipalities. Effective projects require that a special, cross-functional task force or a similar empowered body be put in place. This team should include the representatives of the Department of Public Works (tasked with road engineering, and construction), the municipal utility (in charge of the connection to the power grid and the capacity planning of the electricity), the Information and Communications Technology (ICT) agency (in charge of data networks and cybersecurity), the regional transport authority (ensuring the integration with the public transit), and the urban planning department (to align zoning and the master-plan goals).

The physical change of infrastructure required is significant. In the case of in-road systems, both conductive rail and inductive, a large amount of pavement work is involved. This includes trenching, laying of ERS components and related cabling and repaving. To ensure the long-term viability of such systems and prevent early failure, planners and engineers should carefully examine the structural and thermomechanical implications of such systems embedded in the roadbed, especially asphalt, to ensure the roadside equipment, such as power electronics cabinets transforming grid power and feeding each ERS segment, has space.

The key to the whole system is its connection to the electrical grid. This connection demands the installation of special underground high cables that span between units of closed electrical substations to roadside power cabinets. An important readiness test involves an analysis of the potential of the existing substations that are located along the proposed E-corridors. Very often, these substations will be called upon to carry the large incremental load of the ERS; this investment will have to be planned and funded in very close coordination with the local utility [9].

9.6.3 Link with Sustainable Urban Development Principles

The Electric Road System (ERS) could be used as a powerful tool to promote more comprehensive, sustainable urban development when incorporated in a sensible way. It can also provide the foundation for the development of lively Transit-Oriented Development (TOD) corridors. The ERS enables high-quality, dependable, and zero-emission electric public transport to render residential and commercial districts closer to these lines, which encourages the construction of radically dense, walkable, mixed-use communities to decrease the general dependence on personal vehicles [21].

This adherence to the sustainability principles implies that the deployment of ERS itself will lead to the achievement of several targets in the United Nations Sustainable Development Goal 11 (SDG 11): Sustainable Cities and Communities. Particularly, it facilitates Target 11.2 by availing sustainable transport networks, Target 11.3 by promoting inclusive and sustainable urbanisation, and Target 11.6 by improving the negative environmental effects of cities through better air quality [2].

Physical rollout into the infrastructure of a city compels a resultant and beneficial overlap of long-range planning of three areas that are typically handled individually that is, transport (roads), energy (the grid), and digital (communications networks). Electrification decision on a road corridor should also initiate a similar grid strengthening and data network development plan on the same corridor. This strategy breaks down the departmental silos and initiates a more holistic, efficient and resilient urban development agenda through which the investments in infrastructures are synchronised to attain several goals strategically (Fig. 9.3).

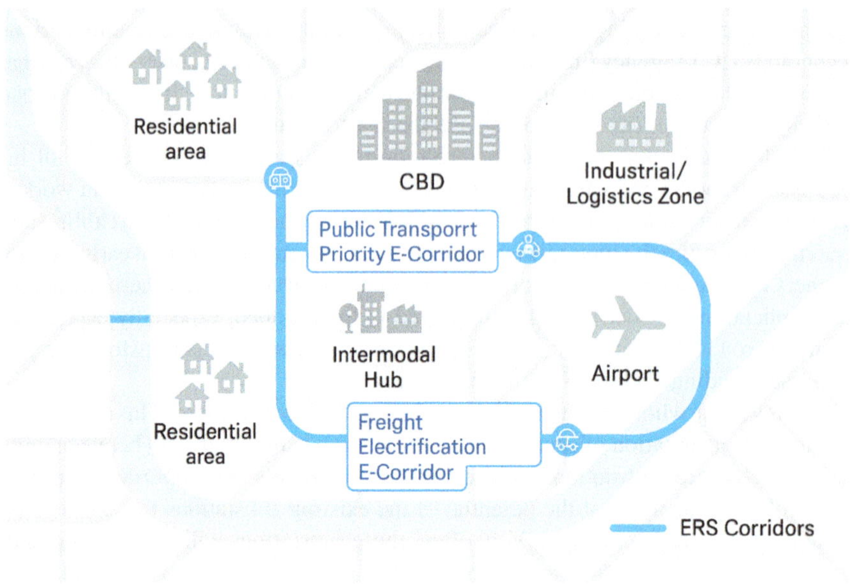

Fig. 9.3 City layout showing ERS-enabled arterial roads and hubs

9.7 Policy and Governance Mechanisms

The implementation of Electric Road Systems at the municipal level will only be successful in the presence of a well-developed system of enabling policies and a well-functioning governance apparatus. The technology provides the tool, but demand, funding and long-term, fair functioning of the system are created by the policy and institutional environment. Therefore, it needs a hybrid system of governance between the regulatory control that is common to a public utility and the operational dynamic of a contemporary technology service provider.

9.7.1 Local Policy Instruments for ERS Adoption

Municipal governments have a range of policy instruments at their disposal to catalyse ERS adoption:

- **Urban EV Mandates**: Fleet electrification mandates are one of the most straightforward methods to establish an initial user base of the ERS. Cities may impose requirements that an increasing fraction of the municipal fleets (e.g., the municipal buses, municipal sanitation fleet, etc.) and commercially licensed fleets (e.g., taxis, ride-hailing services, last-mile delivery vans, etc.) in their cities be zero-emission on certain target dates. These types of mandates will create a predictable and certain demand for charging services, which makes the business case for ERS infrastructure significantly more appealing to investors [22]. The ERS, in its turn, becomes an essential facilitator in these fleets being able to fulfil the requirements without undermining their operational obligations.

- **Green Mobility Incentives**: Cities can supplement their mandates with a package of financial incentives. Although federal incentives usually include tax credits on the acquisition of EVs and charging infrastructure [23], city programs may be more focused. One example of this is local property-tax incentives to logistics firms that operate ERS-compatibility fleets, lowering permit fees on real-estate developments along E-Corridors that install charging infrastructure, or direct subsidies to retrofit the existing commercial vehicles of the company with ERS receivers.
- **Public–Private Partnership (PPP) Models**: Since the infrastructure in ERS is very capital-intensive, PPPs form a fundamental financing and delivery vehicle. Several models can be scaled to the ERS situation:

  Build-Operate-Transfer (BOT): The ERS is financed, designed and built by a private consortium. It then runs the system within a given length of time of concession (e.g., 2030 years), earning usage fees that are paid by vehicle operators. By the expiry of the period, the infrastructure has been transferred to the city [24].
  Availability Payment Model: The city can use the availability payment model in order to de-risk the investment to the private partner. In this system, the state power is paying the commercial operator a fixed and constant fee depending on the presence and functionality of the ERS infrastructure and according to the criteria that have been set regarding the performance (e.g., uptime, power supply). This model will shift the construction and operational risk to the private sector, and the public sector holds the demand and revenue risk [6].

9.7.2 *Institutional Models for Implementation and Operation*

Effective governance also requires establishing the right institutional structures:

- **Smart City Mission Agencies (India)**: In India, the Special Purpose Vehicles (SPVs) that are created through the national Smart Cities Mission can offer a perfect institutional structure to ERS projects. Such SPVs are created to be cross-departmental in nature, fast and nimble, with the responsibility of planning, funding, and delivering complex, technology-focused infrastructure projects, and, therefore, they are well-suited to handle an initiative that falls across the transport, energy, and IT sectors [25].
- **The ERS Operator (ERSO) Role**: There is a risk of the need to develop a new institutional position: the ERS Operator in managing an ERS network. The day-to-day technical operation, maintenance, billing, and customer service of the ERS network would be under the ERSO. This organisation may be a state agency, a commercial organisation on contract, or the PPP consortium. The ERSO would

serve as the key mediator between vehicle users, the infrastructure owner (the city or PPP), and the electric utility, controlling the real-time energy and data flow [26].

9.7.3 Governance Challenges

The implementation of ERS brings with it several critical governance challenges that must be addressed proactively:

- **Interoperability and Standardisation**: The governing body has a major role to play in ensuring that a non-proprietary ecosystem is not created. The policies related to public procurement should require the use of open international standards in all the system elements, such as ISO 15118 in vehicle communication, OCPP 2.0.1 in infrastructure management, and IEC 61850 in grid integration. This guarantees the interoperability and competition between technology providers, and the fact that any vehicle that is compliant will be able to utilise the public infrastructure [27].
- **Data Governance, Privacy, and Security**: ERS networks produce huge amounts of sensitive information, such as vehicle positions, movement history, and energy usage. There should be a data governance framework that is comprehensive. This model should come up with proper guidelines on the ownership of data, access permissions and restrictions of use. It should also have good cybersecurity measures to guard against threats to the system and maintain the privacy of those who will use the system to gain the trust of the people [28].
- **Accountability and Equity**: The form of governance should have good systems of accountability, performance targets of the ERS operator and reporting to the populace. Moreover, it has to take the initiative to solve the problem of social equity. This involves making sure that ERS corridors are laid in a way that cuts across a large section of the community, not only wealthy neighbourhoods or commercial needs and establishing pricing models that are equitable and not an obstacle to low-income consumers (Table 9.3).

9.8 Case Studies

In order to base the conceptual framework of the urban ERS integration on the practical reality, this section examines some of the major pilot projects at the international level. These examples are priceless, data-driven insights into technological capability, performance based on application, governance models and economic issues relating to deployment. Every instance would be analysed with a uniform prism, its technology implemented, the main purpose of its use, the organization structure, and quantifiable results.

Table 9.3 Governance and policy enablers for urban ERS integration

Enabler Type	Specific instrument/ model	Objective
Policy	Municipal fleet electrification mandate	Create guaranteed initial demand and a baseload of users for the ERS network, de-risking the investment
	Streamlined zoning and permitting	Reduce project delays and "soft costs" associated with infrastructure deployment by providing regulatory clarity
Financial	Public–private partnership (PPP)	Leverage private sector capital and operational expertise to finance the high upfront cost of ERS infrastructure
	Local tax credits/fee reductions	Incentivize private fleet operators and real estate developers to adopt ERS technology and build along ERS corridors
Institutional	Smart city agency/SPV	Provide a single, cross-departmental entity with the authority and expertise to manage the complex project implementation
	Mandate for open standards	Ensure long-term interoperability, foster a competitive market for technology, and prevent vendor lock-in
Regulatory	Data governance framework	Establish clear rules for data ownership, privacy, and security to build public trust and ensure responsible data management

9.8.1 Stockholm, Sweden (eRoadArlanda)

- **Technology and Application**: The eRoadArlanda was a project that operated between 2017 and 2019 and involved a 2 km section of conductive rail technology made by Elways/Evias. Its installation in a road surface in a public road between a logistics terminal and the freight terminal at the Stockholm Arlanda Airport was mainly to test and prove the practicality of conductive ground-level charging of commercial freight vehicles such as trucks and buses in practical conditions [29].
- **Governance and Outcomes**: This was a joint initiative by a consortium with the Swedish Transport Administration (Trafikverket) acting as a facilitator, which indicates a partnership between the government and developers of technologies and the end-users of the technology in the logistics industry. The pilot was able to transfer power to levels of 200 kW, and the technology has the potential to be extended to more than 800 kW. It approved the essential safety attribute of dynamic segment activation, where a rail is only energised beneath a compliant vehicle [30]. Although it was technologically successful, the project also revealed the massive financial obstacles of ERS. This pilot was preceded by the cancellation of a larger, permanent 21 km ERS project on the E20 highway scheduled in 2023, which was due to bids being above the set budget, which once again highlights that high capital cost is a major concern to large-scale deployment [31].

9.8.2 *Busan/Gumi, South Korea (OLEV)*

- **Technology and Application**: South Korea has led in the area of wireless ERS, implementing the On-Line Electric Vehicle (OLEV) system created by the Korea Advanced Institute of Science and Technology (KAIST). It is an inductive transfer of power, as in the case of inductive power transfer, the transmitter coils are buried under the road surface, and the receiver pads are mounted under the vehicles to receive the power transmission [32]. Public transport has been the main application, with OLEV-equipped electric buses on the road in some cities since the early 2010s, including Gumi, Sejong, and Daejeon.
- **Governance and Outcomes**: The OLEV projects were developed as government-funded R&D projects by a research institute of the country (KAUST) in cooperation with the local governments. This system demonstrated the feasibility of dynamic wireless charging in public transit so that buses can run with batteries that are about a third the size of those in standard electric buses, thus decreasing the weight and the costs of the vehicle. An important lesson in lifecycle management was also learned in the experience, though: some of the original bus lines became decommissioned because of the high cost and maintenance issues of the ageing first-generation infrastructure. It shows how important it is to consider not only the initial installation cost, but the long-term operational and maintenance costs of the entire business case of ERS [32].

9.8.3 *Berlin/Frankfurt, Germany (eHighway)*

- **Technology and Application**: In Germany, the highway freight sector was the target of the ERS efforts based on the project of the eHighway. The most notable of them is the ELISA (Electrified, Innovative Heavy Traffic on the Autobahn) project on a 10 km stretch of the A5 Autobahn between Frankfurt and Darmstadt, which has been operational since 2019 [33]. It uses conductive overhead lines commonly used with trains and trolleybuses. Hybrid trucks are specially equipped with a pantograph, which connects them to the lines and extracts power when travelling on the highway, helping them to run on electricity and at the same time charging their batteries to compensate for the non-electrified parts of their travel [33].
- **Governance and Outcomes**: The ELISA project, a federally funded project managed by the state of Hessen, is based on a partnership that involves Siemens as technology provider and a group of logistics companies as end-users. The pilot phase has had empirical results that demonstrate that the catenary system has an efficiency of about twice that of a conventional internal combustion engine. Quantitative calculations show that a 40-ton truck can save EUR 16,000 in fuel costs when the transport is used to go over a distance of 100,000 km. The trial has also confirmed the strength and stability of the catenary technology under intense and realistic conditions on the highway. The evidence provided in the case studies

highlights a rather unique pattern of application-technology fit: high-power, high-efficiency catenary systems prove to exhibit the best performance when inter-city heavy freight must be transported over major highways, when the visual impact can be considered relatively insignificant. On the other hand, in-ground systems (inductive systems and rail systems) are currently being investigated in intra-city applications like bus fleets and urban logistics, where overhead lines are not so feasible due to their aesthetic imperatives and the complexity of urban streets. These observations imply that municipal authorities must not focus on a single optimal ERS technology per se; instead, they ought to follow the technology that better fits the unique urban mobility problem before them.

9.8.4 Potential in Pune or Bengaluru, India

- **Context and Opportunity**: Pune and Bengaluru are the main cities that are engaged in the program of Smart Cities in India, and have a high goal of adopting electric vehicles. Pune already has one of the biggest electric bus fleets that the city operates under its public transport utility, PMPML [34]. Bengaluru is a tech dynamic and movement hub, and it is estimated that more than 2.3 million electric vehicles will be on its streets by 2030 [35]. Such urban settings are a good breeding ground for a hardcore ERS pilot.
- **Potential Application**: An interesting pilot project would be the electrification of a special Bus Rapid Transit (BRT) line in Pune or one of the major arterial roads in Bengaluru that links the vast technology parks in the city. This project would correspond to the current e-mobility projects and could be managed through the municipal Smart City Special Purpose Vehicle (SPV) framework, which is intended to support cross-sectoral projects [25]. The Indian pilot would give priceless empirical information on ERS performance, cost, and operational issues in the special context of a large, fast-growing emerging economy (Fig. 9.4 and Table 9.4).

9.9 Future Smart City Framework for Asia

Although the experience of European and North American ERS pilots can help to provide valuable technical knowledge, a simple approach of copy-pasting will not work in the environment of heterogeneously presented Asian megacities. Such urban spaces are often typified by high population densities, a more heterogeneous and anarchic traffic composition including a high share of two and three-wheelers, different institutional and regulatory topographies and in many instances the possibility of bypassing legacy infrastructure with next-generation digital infrastructure. Therefore, an ERS integration framework to be developed in the future to serve the Asian context will have to be adapted to these realities, with the advanced digital technology utilised to provide complex management and resilience. An improved

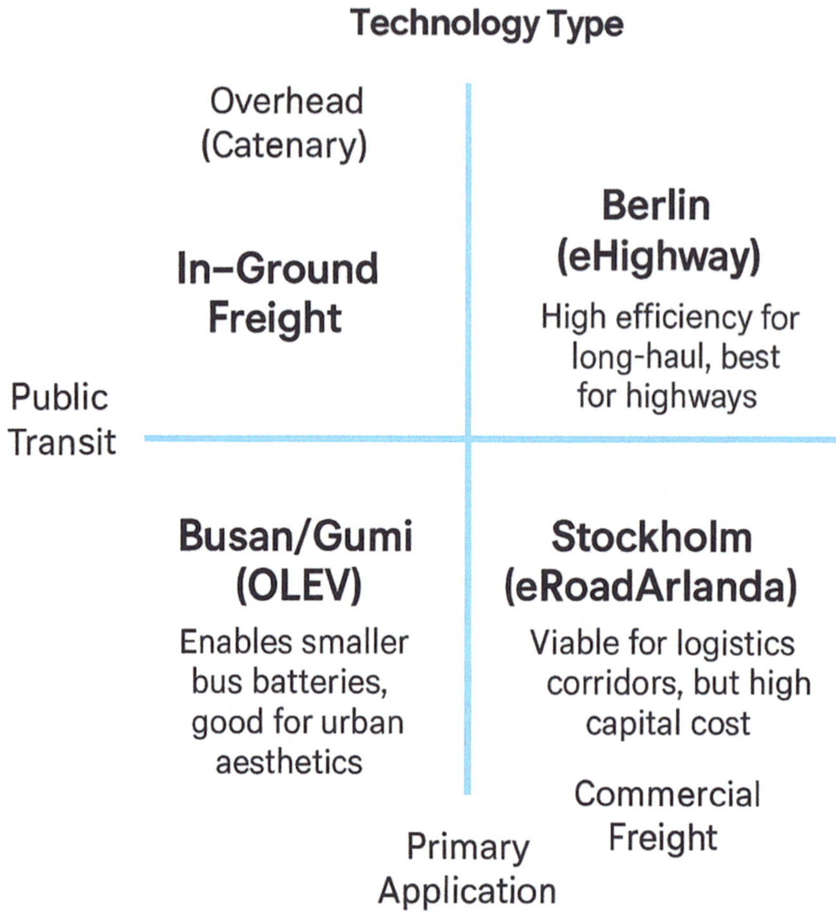

Fig. 9.4 Comparative case summary schematic

Asian smart city is proposed within the conceptual model below, which is based on four integrated pillars.

9.9.1 Pillars of the Conceptual Model

1. **Digital Twin Infrastructure Management**: In the model, the ERS is not perceived as a frozen physical construction, but may be considered as a dynamic part of a city-wide digital twin. It involves the development of a comprehensive virtual representation of the physical infrastructure of the ERS based on Building Information modelling (BIM) data on its physical objects (rails, coils, power cabinets) and combining it with a continuous flow of real-time Internet of

Table 9.4 City case comparison

City	Country	System type	Energy transfer	Route length	Primary application	Data integration	Key outcomes/lessons learned
Stockholm (Arlanda)	Sweden	Conductive rail	Conductive (in-road)	2 km	Commercial freight	Vehicle identification and billing system	Technologically feasible for logistics; high capital cost remains a major barrier for large-scale deployment
Busan/Gumi	South Korea	Inductive	Wireless (in-road)	Various	Public buses	Operational control and monitoring systems	Enables smaller bus batteries, improving vehicle efficiency; long-term maintenance and lifecycle costs are critical challenges
Berlin (A5)	Germany	Catenary	Conductive (overhead)	10 km	Long-haul freight	Vehicle-pantograph communication, energy monitoring	Highly efficient and robust for heavy trucks on highways; visual impact may limit its use in dense urban centers
Pune (Proposed)	India	Inductive/rail	TBD	TBD	Public buses (BRT)	Integration with smart city command and control center	Potential to solve e-bus charging challenges in dense, high-traffic emerging economies; a critical test for scalability

Things (IoT) data on its working conditions (power flow, temperature, vehicle traffic) [12]. A living model can be used to transition between reactive and predictive maintenance so that the operator can detect and remove possible problems, like a degraded power segment or overheating transformer, before the system fails. It can also be an effective smart city simulation to help the urban planners model the effects of an increasing ERS or changes in traffic.

2. **City Energy Command Centre (ECC)**: The whole of the ERS network would be operated as an important controllable property in a centralised City Energy Command Centre. This ECC would act as a data-driven hub by not only incorporating real-time data in the ERS, but also building energy management systems, distributed renewable sources of energy (e.g., rooftop solar), battery storage systems and the overall power grid [36]. The ECC would be able to constantly optimise energy flows within the city by means of artificial intelligence (AI) and machine learning. As an example, it may automatically raise power supply to the ERS and provide lower prices to charge at times of excess solar production and use the linked vehicles' fleet to even out grid needs [37].
3. **AI-Based Traffic and Energy Synchronisation**: This pillar forms the active centre of the framework, where mobility and energy control are very closely synchronised using an AI-based optimisation platform. The platform receives the information provided by the ECC (grid condition, energy costs) and the Traffic Management Centre (congestion, traffic flow) and implements complex, coordinated measures. As an illustration, in case the ECC is sending warnings of grid stress, the AI platform can also communicate with the traffic signal system to reterritorialize vehicles that might be looking to enter the most energy-demanding ERS areas. On the other hand, to take in an unexpected burst of renewable energy, it might coordinate a green wave of traffic lights along a particular electric-powered highway, persuading electric vehicles to travel the highway and recharge [38]. This change makes the ERS a dynamic load-shaping and traffic management tool and not a passive energy consumer.
4. **Dynamic Pricing and Local Energy Markets**: The framework suggests the departure of the inactive, per-kilowatt-hour pricing system of ERS use. In its place, there would be a dynamic pricing system whereby the price of accessing energy at the ERS is adjusted dynamically based on grid conditions, congestion, and time of day. These market-based incentives would encourage drivers and fleet operators to make behavioural changes that would allow the entire system to be efficient.

9.9.2 Linkages to National Digital Transformation Agendas

This advanced framework is not a standalone vision but is designed to align with and accelerate existing national digital transformation initiatives across Asia:

- **Digital India**: The data-centricity of the proposed model is closely aligned with the objectives of the Digital India program, which seeks to establish a strong digital infrastructure (including the BharatNet rural broadband project) and create a broad array of digital services [39]. The ERS data platform and the ECC would be eager to integrate directly into the Integrated Command and Control Centres (ICCCs) being developed under the Smart Cities Mission of India to create a strong digital infrastructure.

- **Thailand 4.0**: This is a national strategy that is expressly aimed at transforming the Thai economy to a high-value, technologically advanced industry, such as Next-Gen Automotive and Digital Economy [40]. An ERS implementation, specifically with the advanced AI and digital twin concepts of this framework, would be a flagship project of Thailand 4.0 that would catalyse local R&D and manufacturing in these specific target sectors.
- **Indonesia Smart City Master Plan**: The national smart city plan of Indonesia is developed on six dimensions, featuring Smart Mobility, Smart Environment and Smart Governance, but is largely based on the power of ICT [41]. The offered ERS framework is aimed at all these dimensions. Moreover, the implementation of it may be directed and quantified by the national standard of smart city in Indonesia, SNI ISO 37122, which offers the official set of indicators to evaluate the performance in the spheres of energy, transport, and ICT [42].

To most Asian megacities that are grappling with the dual realities of hyper-congestion and cramped power grids, the core benefit of an intelligent ERS can be further than decarbonization. It is a fact that the system provides an effective, powerful tool for AI-driven management of these extreme conditions, which are dynamic. The system enables advanced trade-offs and optimisations to be made between real-time grid data and real-time traffic data by establishing an intelligent connection between the two, which can improve the overall resilience of the city. Here, ERS is not simply an environmental solution; it is a strategic tool for handling the main complexities of the operations of the city of the twenty-first century, in Asia.

Figure 9.5 would be a comprehensive, multi-layered schematic diagram illustrating the proposed framework.

- The **Physical Layer** at the bottom would show ERS-enabled roads with icons for buses, trucks, and cars, connected to a smart grid with renewable energy sources (solar, wind) and energy storage.
- The **Data and Communications Layer** above it would depict a network of IoT sensors on the roads and vehicles, communicating via 5G/C-V2X to a central cloud platform. This platform would be labeled "City Digital Twin," showing it integrates BIM models of infrastructure with real-time IoT data.
- The **Control and Optimization Layer** would be the core of the diagram. It would feature two main components: a "City Energy Command Center (ECC)" and a "Traffic Management Center (TMC)." Between them, a central "AI Synchronization Engine" would be shown, receiving data from both and sending control signals back to them. Arrows would indicate that the ECC manages grid balance and renewable energy integration, while the TMC manages traffic signals and vehicle routing.

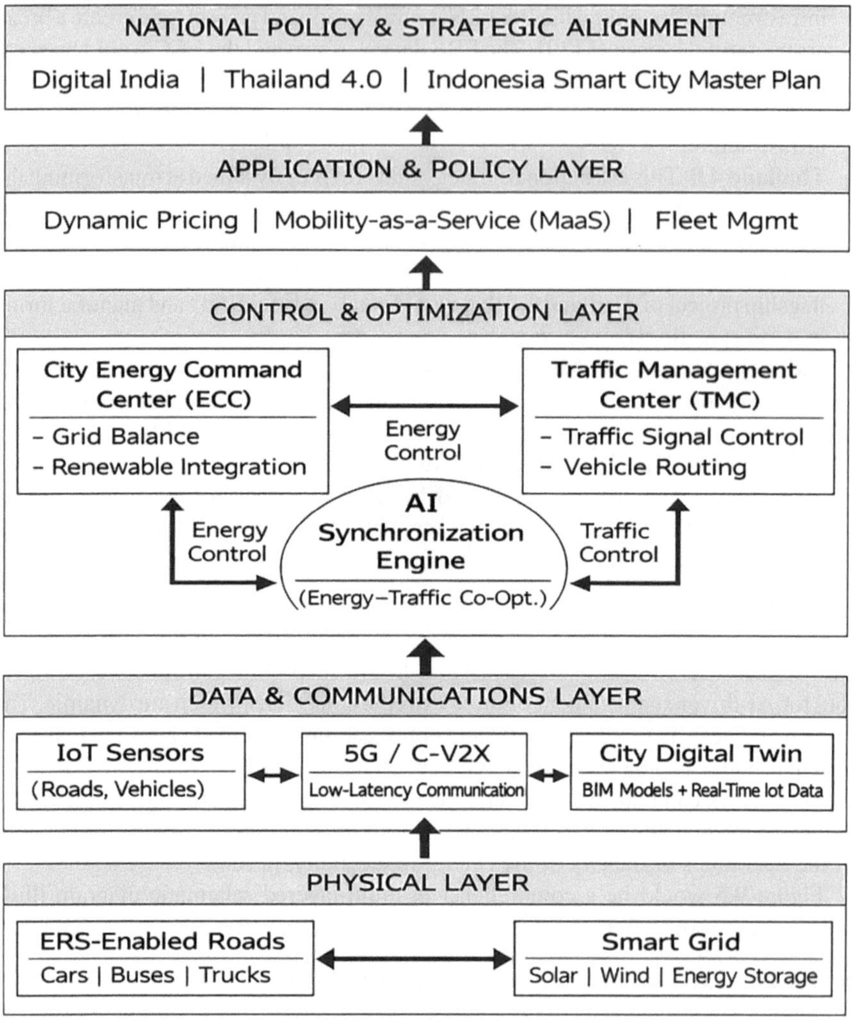

Fig. 9.5 Conceptual blueprint of ERS-integrated smart city for Asia

- The **Application and Policy Layer** at the top would show the outputs and guiding inputs. Icons for "Dynamic Pricing," "Mobility-as-a-Service Apps," and "Fleet Management Dashboards" would represent the user-facing services. Above all, overarching banners labeled with national initiatives like "Digital India," "Thailand 4.0," and "Indonesia Smart City Master Plan" would indicate that the entire framework is aligned with and driven by national policy objectives. The diagram would clearly communicate a system-of-systems approach, where physical infrastructure is managed through a data-driven, AI-powered control layer to achieve strategic policy goals.

9.10 Challenges and Opportunities

Although the development of Electric Road Systems in smart cities provides a revolutionary perspective of mobility and energy control within cities, it is full of serious technical, economic, social, and environmental obstacles on the way to the mass frontier. An objective evaluation of such obstacles is essential in coming up with mitigation measures. Every difficulty, however, has an innovation and a strategy to consider. The most significant non-financial obstacle is, perhaps, the institutional inertia because the technology will require a degree of cross-departmental cooperation, which not all city governments are designed to be naturally.

9.10.1 Technical Challenges

- **Interoperability**: One of the major technical risks is the spread of proprietary non-standardised ERS technologies. When cities or regions move to closed ecosystems, then there may be lock-in on vendors, with only certain models of vehicles able to access the infrastructure, which would then divide the market and reduce the network effect. This poses a big challenge to the establishment of a smooth, accommodating system for everyone [27].
- **Cybersecurity**: A smart ERS is based on an extensive V2I communication network that is a large attack surface. There is a potential risk of malicious actor exploitation of the vulnerabilities to interfere with the traffic flow, destabilise the power grid with the aid of manipulation of charging loads, steal sensitive user data, and even jeopardise the safety of vehicles. The security and resilience of this multi-faceted cyber-physical setup are the most critical issue [43].
- **Legacy System Integration**: Most cities are overrun with various systems of legacy IT and operational technologies. The technical challenges of integrating a state-of-the-art ERS control platform with older municipal traffic management software, ageing utility SCADA systems and disparate billing platforms may be difficult, time-consuming and costly.

9.10.2 Economic Challenges

- **High Capital Costs**: The greatest obstacle to the ERS implementation is the high upfront capital cost. The infrastructure installation estimates are very varied in cost, depending on the technology, local conditions, but are always between EUR 1.7 and EUR 3.1 million per kilometre to install a modest network of the main corridors of the city. This becomes a significant investment that may be hard to finance with the regular municipal budgets.
- **Long and Uncertain Payback Periods**: In the case of a public authority, ERS has a complicated payback period that can take a number of years. The benefits are

both direct (usage fees) as well as indirect (societal benefits such as lower healthcare expenses due to cleaner air and economic benefits due to less congestion), which may be hard to measure and explain politically. For this reason, it is difficult to attract private investment in the absence of substantial public de-risking arrangements like the PPPs.

Social Challenges

- **Public Awareness and Acceptance**: Mega-infrastructure development inevitably brings havoc. The erection process of ERS, which includes massive road activities, can cause societal resistance because of the traffic rerouting and the noise. It is also possible that there are popular issues of safety of in-road electrical systems or simply the aesthetics of overhead catenary lines, and this can only be overcome through proactive and open public awareness and involvement campaigns, which clearly explain the long-term benefits [44].
- **Equity of Access**: One of the key social risks is the possibility of inequitable distribution of the benefits of ERS. When E-Corridors are constructed in the wealthier business areas or cater to only the market of high-end logistics businesses, then it may widen the mobility gap and leave the poorer communities underserved. It needs to be planned and a policy taken to ensure that ERS implementation, especially in the context of public transport, focuses on serving a wide variety of neighbourhoods [28].

Environmental Challenges

- **Construction Footprint**: Although ERS has extensive operational environmental advantages of decreasing tailpipe emissions, its construction phase has an environmental impact. Embodied carbon and resource use in the manufacturing of steel, copper, and electronics, as well as during the installation process that includes the use of concrete and asphalt, have to be included in a full lifecycle assessment.
- **Pavement Durability**: Charging infrastructure added on the vehicular roadways presents new failure modes. The mechanical stresses and the differential thermal expansion existing between the electric road system (ERS) hardware and the ambient pavement composition might accelerate the degradation of the material, leading to the crack initiation and development of ruts. Subsequently, there is an increased potential for maintenance rates and costs in the run-time life cycle of the system.

The triumphant clearing of these obstacles does not hinge as heavily on a technological breakthrough applied individually and as much on the ability of a municipality to undergo institutional reorganisation. Although technical barriers can be allayed by the introduction of industry standards and the economic barriers can be handled by creative financing vehicles, the actual implementation of them necessitates a highly concerted action within the various departments of planning, transport, energy, and information technology. Within a traditional isolating municipal organisation with individual departments and different budgets, priorities, and horizons in planning, the achievement of the necessary degree of synchronisation is a daunting governance

Table 9.5 Challenges and potential policy/technical mitigations

Challenge category	Specific challenge	Potential mitigation strategy
Technical	Lack of interoperability	Mandate adherence to open international standards (ISO 15118, OCPP 2.0.1, IEC 61850) in all public procurement contracts for ERS
Technical	Cybersecurity vulnerabilities	Implement a defense-in-depth security architecture based on a recognized framework (e.g., NIST); require end-to-end encryption and regular third-party penetration testing
Economic	High capital expenditure	Utilize public–private partnership (PPP) models (e.g., availability payment) to leverage private financing; actively seek national and international green infrastructure funds
Economic	Uncertain business case	Focus initial deployment on high-utilization commercial and public fleets (buses, freight) where the ROI is clearest and fastest, creating a foundational user base
Social	Public opposition to construction	Develop a phased deployment plan to minimize disruption in any single area; launch sustained public awareness campaigns highlighting long-term benefits like cleaner air and reduced congestion
Social	Equity concerns	Prioritize ERS deployment on public transport corridors that serve diverse and underserved communities; use a portion of ERS revenue to subsidize electric mobility options for low-income groups
Environmental	Pavement degradation	Invest in R&D for advanced, durable pavement materials specifically designed to integrate ERS components; utilize digital twin models for predictive maintenance of road surfaces

challenge. In this regard, the creation of an independent, empowered, cross-cutting institution, including a Smart City special purpose vehicle (SPV) or an e-corridor authority, is often a precondition towards success and makes the ERS initiative less of a trial of new technology but rather a test of the ability of the city to govern itself in an integrated and modern way (Table 9.5).

9.11 Conclusion

As has been shown in this chapter, the introduction of the Electric Road Systems (ERS) into the urban environments is a strategic development that cannot be limited to a simple vehicle charging system. ERS is a paradigm of transformative infrastructure that can be viewed as an effective driving force of more intensive integration of smart-city paradigms, which physically and digitally connect the previously separate domains of mobility, energy, data, and governance. Allowing moving, in-motion charging, ERS essentially transforms the economic and logistical aspects of city fleet electrification, which concern the challenges of grid overload, battery-reliance issues, and air-pollution goals in a holistic, systemic way.

The achievement of this vision would be possible due to the presence of a twig of essential enablers that were identified and analysed. Technologically, the combination of the Internet of Things, edge computing, and artificial intelligence is vital towards creating a responsive, predictive, and optimised system. In terms of standards, strict compliance with an interoperable protocol bridge, which combines automotive standards (ISO 15118), EV-charging standards (OCPP 2.0.1), and power-grid standards (IEC 61850), is a prerequisite to have an open, competitive, and future-proof ecosystem.

At the institutional level, the high cost and cross-cutting of ERS requires new models of governance, especially the use of the concept of public–private partnership to mobilise private resources and the formulation of empowered, cross-departmental agencies like Smart City organisations to overcome the municipal silos. Finally, this technological and institutional structure should be anchored on a long-term vision entrenched in the urban planning in which the basic fabric of a sustainable, transit-oriented city is integrated within the city master plans as the engine of ERS corridors.

The future of the ERS-enabled smart city, particularly in the rapidly urbanising and technologically dynamic regions of Asia, holds immense potential. By moving beyond the models of early pilots and embracing a deeply integrated, data-driven approach, cities can leverage ERS not just for decarbonization but as a strategic tool for managing the complex interplay of traffic congestion and energy demand, thereby building more resilient, efficient, and livable urban futures. While this chapter has laid the groundwork for the systemic integration of ERS, the full potential of this infrastructure will be unlocked by next-generation technologies. The next chapter, which is to be called AI and Emerging Technologies in ERS, will discuss how improvements in artificial intelligence, quantum computing, and new materials can lead to further revolution of the capacity, efficiency and intelligence of Electric Road Systems.

References

1. f3 Centre. Electric road systems—fact sheet (Misc-4_210510). https://f3centre.se/app/uploads/Electric-Road-Systems-fact-sheet_Misc-4_210510.pdf. Accessed 18 Oct 2025
2. United Nations Environment Programme (UNEP). Goal 11: sustainable cities and communities. https://www.unep.org/topics/sustainable-development-goals/why-do-sustainable-development-goals-matter/goal-11-0. Accessed 18 Oct 2025
3. MDPI. Benefits of an electric road system for battery electric vehicles. https://www.mdpi.com/2032-6653/13/11/197. Accessed 18 Oct 2025

4. ResearchGate. Integrative ITS platform for enhanced urban mobility in smart cities. https://www.researchgate.net/publication/393548383_Integrative_ITS_Platform_for_Enhanced_Urban_Mobility_in_Smart_Cities. Accessed 18 Oct 2025
5. IBM. What is a smart city? https://www.ibm.com/think/topics/smart-city. Accessed 18 Oct 2025
6. U.S. Department of Treasury. Expanding the market for infrastructure public–private partnerships: alternative risk and profit-sharing approaches. https://home.treasury.gov/system/files/226/Expanding_the_Market_for_Infrastructure_Public_Private_Partnerships_Alternative_Risk_and_Profit_Sharing_Approaches_to_Align_Sponsor_and_Investor_Interests_APR2015.pdf. Accessed 18 Oct 2025
7. PubMed Central (PMC). Integration of IoT-enabled technologies and AI for smart city scenarios: recent advancements and future trends. https://pmc.ncbi.nlm.nih.gov/articles/PMC10256108/. Accessed 18 Oct 2025
8. Cavli Wireless. How IoT in smart cities powers urban transformation: architecture, benefits, and key solutions. https://www.cavliwireless.com/blog/nerdiest-of-things/iot-smart-city-solutions. Accessed 18 Oct 2025
9. Futura-Mobility. Electric road systems: towards decarbonising road transport. https://futuramobility.org/en/%F0%9F%93%8C-electric-road-systems-towards-decarbonising-road-transport/. Accessed 18 Oct 2025
10. Federal Highway Administration (FHWA). Congestion pricing: environmental benefits. https://ops.fhwa.dot.gov/congestionpricing/resources/enviro_benefits.htm. Accessed 18 Oct 2025
11. MDPI. Impact of transportation electrification on the electricity grid—a review. https://www.mdpi.com/2624-8921/4/4/56. Accessed 18 Oct 2025
12. Blyncsy. Digital twins for road networks drive proactive change. https://blyncsy.com/ai/digital-twins-for-road-networks/. Accessed 18 Oct 2025
13. MDPI. An IoT architecture for sustainable urban mobility: towards energy-aware and low-emission smart cities. https://www.mdpi.com/1999-5903/17/10/457. Accessed 18 Oct 2025
14. MDPI. Integration and exploitation of sensor data in smart cities through event-driven applications. https://www.mdpi.com/1424-8220/19/6/1372. Accessed 18 Oct 2025
15. Driivz. What is ISO 15118? https://driivz.com/glossary/iso-15118/. Accessed 18 Oct 2025
16. Open Charge Alliance. OCPP (open charge point protocol). https://openchargealliance.org/protocols/open-charge-point-protocol/. Accessed 18 Oct 2025
17. ResearchGate. Integrating IoT and blockchain for smart urban energy management: enhancing sustainability through real-time monitoring and optimization. https://www.researchgate.net/publication/396372350_Integrating_IoT_and_blockchain_for_smart_urban_energy_management_enhancing_sustainability_through_real-time_monitoring_and_optimization. Accessed 18 Oct 2025
18. WiPowerOne. E-Bus—wireless EV charging. https://www.wipowerone.com/eng/business/bus.php. Accessed 18 Oct 2025
19. Taylor & Francis Online. Micromobility for first and last mile access to public transport. https://doi.org/10.1080/07293682.2023.2211690. Accessed 18 Oct 2025
20. Alternative Fuels Data Center (AFDC). Building codes, parking ordinances, and zoning ordinances for electric vehicle charging infrastructure. https://afdc.energy.gov/fuels/electricity-codes-and-ordinances. Accessed 18 Oct 2025
21. The Energy and Resources Institute (TERI). Integrating climate sensitivity into urban planning for future resilience. https://teriin.org/sites/default/files/files/Integrating_Climate_Sensitivity_into_Urban_Planning_for_Future_Resilience.pdf. Accessed 18 Oct 2025
22. Sustainability Directory. EV adoption mandates. https://energy.sustainability-directory.com/term/ev-adoption-mandates/. Accessed 18 Oct 2025
23. Alternative Fuels Data Center (AFDC). Tax credits for electric vehicles. https://afdc.energy.gov/laws/ev-tax-credits. Accessed 18 Oct 2025
24. National Bureau of Economic Research (NBER). When and how to use public–private partnerships in infrastructure. https://www.nber.org/system/files/working_papers/w26766/w26766.pdf. Accessed 18 Oct 2025

25. Smart City Trivandrum. About 'Smart City Mission'. https://smartcitytvm.in/index.php/home/about-smart-city-mission. Accessed 18 Oct 2025
26. DiVA Portal. Research and innovation platform for electric road systems. https://www.diva-portal.org/smash/get/diva2:1589096/FULLTEXT01.pdf. Accessed 18 Oct 2025
27. PIARC. Electric road systems: a solution for the future. https://www.piarc.org/en/order-library/29690-en-Electric%20road%20systems:%20a%20solution%20for%20the%20future. Accessed 18 Oct 2025
28. Prism Sustainability Directory. Smart city data governance frameworks. https://prism.sustainability-directory.com/scenario/smart-city-data-governance-frameworks/. Accessed 18 Oct 2025
29. f3 Centre. Electric road systems (ERS). https://f3centre.se/en/fact-sheets/electric-road-systems-ers/. Accessed 18 Oct 2025
30. Wikipedia. Swedish transport administration electric road program. https://en.wikipedia.org/wiki/Swedish_Transport_Administration_electric_road_program. Accessed 18 Oct 2025
31. Trafikverket Bransch. Electric road E20, Hallsberg–Örebro. https://bransch.trafikverket.se/en/startpage/projects/Road-construction-projects/electric-road-e20-hallsbergorebro/. Accessed 18 Oct 2025
32. Wikipedia. Online electric vehicle. https://en.wikipedia.org/wiki/Online_electric_vehicle. Accessed 18 Oct 2025
33. Merck Group. Driving the electric highway. https://www.emdgroup.com/en/research/science-space/envisioning-tomorrow/scarcity-of-resources/e-mobility-hot-topic.html. Accessed 18 Oct 2025
34. NITI Aayog. E-mobility division. https://niti.gov.in/divisions/division/e-mobility. Accessed 18 Oct 2025
35. Center for Study of Science, Technology and Policy (CSTEP). Bengaluru 2030: EV charging demand and infrastructure. https://cstep.in/publications-details.php?id=3004. Accessed 18 Oct 2025
36. Schneider Electric. Energy command center (ECC). https://www.se.com/ww/en/work/solutions/power-management/energy-command-center/. Accessed 18 Oct 2025
37. Smart Buildings Magazine. The energy command center: the key to AI-powered, net-zero cities. https://smartbuildingsmagazine.com/features/the-energy-command-center-the-key-to-ai-powered-net-zero-cities. Accessed 18 Oct 2025
38. SWARCO. AI-based traffic management. https://www.swarco.com/stories/ai-based-traffic-management. Accessed 18 Oct 2025
39. Press Information Bureau (PIB). Empowering governance through digital infrastructure. https://www.pib.gov.in/FactsheetDetails.aspx?Id=149256. Accessed 18 Oct 2025
40. Emerhub. What the Thailand 4.0 initiative means for your business. https://emerhub.com/thailand/what-the-thailand-4-0-initiative-means-for-your-business/. Accessed 18 Oct 2025
41. ResearchGate. The strategy smart city development concepts in Indonesia. https://www.researchgate.net/publication/360083194_The_Strategy_Smart_City_Development_Concepts_in_Indonesia. Accessed 18 Oct 2025
42. Asia-Pacific Economic Cooperation (APEC). Implementation of smart city standards in Indonesia based on SNI. https://www.apec.org/docs/default-source/publications/2022/6/implementation-of-smart-city-standards-in-indonesia/222_scsc_implementation-of-smart-city-standards-in-indonesia.pdf. Accessed 18 Oct 2025
43. MDPI. Threat landscape and integrated cybersecurity framework for V2V and autonomous electric vehicles. https://www.mdpi.com/2032-6653/16/8/469. Accessed 18 Oct 2025
44. Drive Electric Campaign. Home. https://www.driveelectriccampaign.org/. Accessed 18 Oct 2025

Chapter 10
Conclusions and Future Research Directions: Charting the Path to a Systemic Mobility Transition

10.1 Overview of Key Findings: A System-Of-Systems Perspective

In the previous chapters of this monograph, the technological, infrastructural, economic and socio-political aspects of Electric Road Systems (ERS) have been deconstructed in a systematic way. This conclusive chapter would combine all these multifaceted analyses into a consistent forward-looking narrative, to the extent that ERS is not just an alternative system of charging electric vehicles, but a paradigm shift in mobility infrastructure. ERS is a classic system-of-systems, a nexus technology that essentially operates in the transport, energy and digital domains, providing a systemic solution to the omnipresent issue of decarbonisation. This synthesising approach demonstrates that ERS is not in its individual parts, but in its ability to make vehicles, infrastructure, and the energy grid symbiotic with each other, and therefore unlock efficiencies and capabilities that are not enabled with siloed methods.

The main thesis of the work lies in the fact that the worldwide shift to sustainable transport needs to shift away from a paradigm based on subsidising the purchase of individual vehicles to one that prioritises infrastructures, which could allow new forms of operation and new technologies of vehicles. Initial policies towards electric vehicles (EVs), like the early stages of the FAME scheme in India, and other national programmes in the European Union, were characterised by much emphasis on demand-side incentives.1 Although the initial policy was initially effective at seeding the market, it created a long-standing chicken-and-egg problem, with mass uptake of the technology inhibited by infrastructure anxiety and the constrained nature of infrastructure investment by a small number of vehicles. ERS gets out of this dilemma by its very nature; it is an infrastructure-based solution that allows implementing a new type of EVs, the ones with much smaller batteries, and allows the implementation of new forms of operation like continuous freight logistics. Recent significant policy shifts, such as the National Electric Vehicle Infrastructure (NEVI) Programme of the United States and the requirement of millions of public charging

R. Agnihotri, *Electrified Road Systems (ERS) for Emerging Asia*, Springer Tracts on Transportation and Traffic 24, https://doi.org/10.1007/978-3-032-16927-3_10

points in the EU, are indicators of a strategic shift towards the infrastructure-first logic, which can be overcome by the vehicle-centric policy.

This systemic integration is driven by a confluence of key innovations that have been explored throughout this monograph.

- **Dynamic Wireless Power Transfer (DWPT)**: Based on the discussion in Chaps. 6 and 7, DWPT is an innovative technology that enables vehicles to charge on the move, virtually eliminating range anxiety and allowing the on-board battery size to be greatly reduced. The implications of this are far-reaching in terms of the economic feasibility, through the reduced cost of the vehicles, and resource sustainability, because this will reduce the strain on key battery minerals. The main pillar of the ERS value proposition is its capability to transform the logistics of urban areas and long-haul freight management by reducing downtime.
- **Smart Grid Integration**: ERS reinvents the connection between the transportation and the power grid. Instead of a huge, inactive user of electricity, an electrified road network is an active and movable grid resource. By supporting smart grid technologies, ERS may be involved in dynamic load management, where vehicle charging demand is shifted to off-peak time or during periods of heavy renewable energy production [8]. In more sophisticated forms, vehicle-to-grid (V2G) can enable the overall scalability of battery capacity of vehicles on the network to be used to provide ancillary services, improving grid resilience, and enabling a greater penetration of intermittent renewable energy [10].
- **AI-Based Control Systems**: Due to the complexity of running a large-scale ERS network, there is a need to move away from passive to intelligent management. The algorithms of machine learning and Artificial Intelligence (AI), as will be discussed in Chap. 8, are important to optimise the system in real-time. Such digital systems control the dynamic dispatch of energy to vehicles in motion, optimise traffic flow to avoid congestion, support vehicle-everywhere (V2X) communications with vehicles to achieve a higher level of safety and predictive maintenance of the infrastructure itself, making roadways responsive and intelligent at the same time [12].
- **Sustainable Materials and Circular Economy Principles**: The ERS as a physical infrastructure should be sustainable itself. One of the findings of this monograph is the relevance of the life-cycle approach to ERS construction and maintenance. This includes the utilisation of greener materials, including recycled and permeable asphalt, low-carbon concrete, and reclaimed industrial by-products [14], and circularly designing ERS infrastructure such that it can be recovered and recycled in the future, including embedded coils, power electronics, and paving materials, is a critical part of minimising its long-term environmental footprint and is consistent with the principles of a circular economy [18].

These interconnected innovations are best visualised as an integrated system, as illustrated in Fig. 10.1.

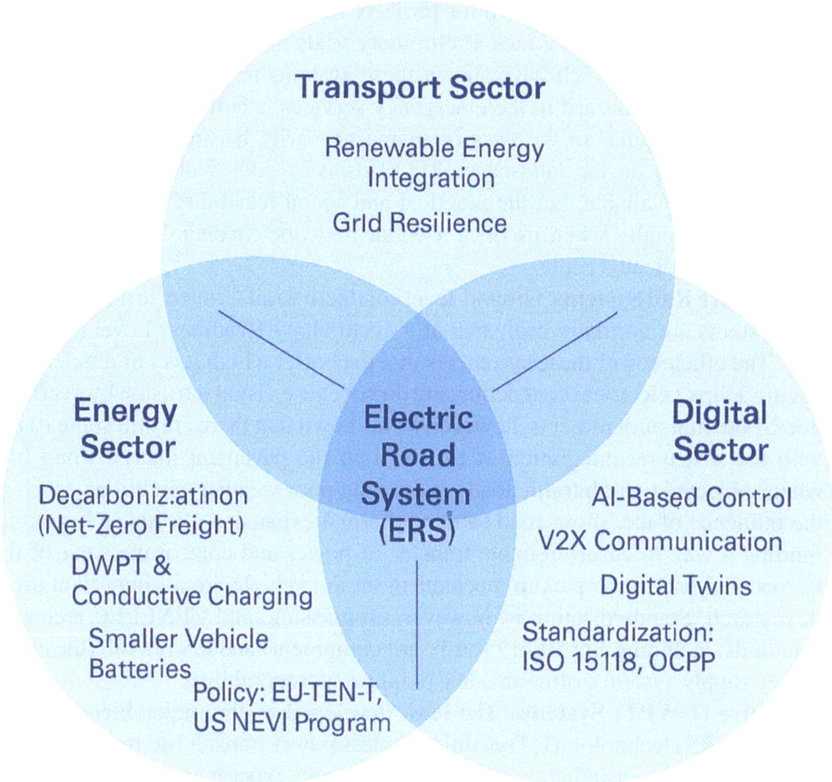

Fig. 10.1 ERS technology-policy integration map

10.2 Technological Evolution and Maturity: From Pilot to Practice

Although the ERS conceptual framework is quite strong, its practical implementation depends on the maturity and preparedness of its central technologies. The Technology Readiness Level (TRL) analysis shows a complex situation in which technical maturity does not necessarily translate into deployment readiness. The most technologically advanced systems have high practical barriers, whereas less advanced technologies can have higher potential in the long-term perspective because of their high level of flexibility and acceptance among the population.

A review of global pilot projects and technical assessments provides a nuanced picture of the current state of ERS technology as of 2024.

- **Overhead Catenary Systems**: Overhead line systems are the most developed ERS solution, which have been historically developed over several decades in the rail sector, thereby able to provide high power levels (up to 98%) and apply to heavy trucks. Nonetheless, pilot projects in Europe have recently revealed quite serious deployment issues. Preliminary trials in Germany and France have reported final tests which have shown these systems to be costly to install and maintain, a safety hazard to the emergency services, a collision risk to roadside poles, and unpopular in the eyes of many, primarily because of their adverse aesthetic effects on the landscape [21]. This is at odds with the fact that the technology is available, but the practical and social feasibility of the systems is doubtful, especially when used on a wider network coverage rather than on a limited and dedicated route.

- **Conductive Rail Systems**: Ground-level conductive rail Ground-level conductive rail systems are normally evaluated at a Technology Readiness Level (TRL) of 5–7. The efficiency of these systems is that they offer a high level of efficiency in having a direct electrical connection and do not cause visual intrusion by overhead lines. Potential pilot projects, however, have shown that there are still some issues with the structural integration of the rail into the pavement matrix. Long life, when subjected to high traffic loads and during poor weather conditions - such as the influence of ice, snow, road salt and thermal expansion—is also a key issue. Finding a way to ensure reliable transfer of power and control the wear of the in-road rail and in the pickup mechanism on the vehicle are all important areas of research. Standardisation is, however, progressing, and CENELEC technical standards, including EN 50,717 (on-board equipment) and EN 50,740 (the entire power-supply system), offer an entry point to interoperability.
- **Inductive (DWPT) Systems**: The least developed of the major Electric Road System (ERS) technologies, Dynamic wireless power transfer has the majority of evaluated system installations at TRL 4–6. Recent experimental outcomes have been pointing to major obstacles. A power-transfer efficiency of 64.3% in a 2023 German trial and significant installation challenges. The thermomechanical stress caused by the heat produced by inductive coils on the materials that constitute the conventional roads has been identified as a source of French investigations, which has heightened the chances of degrading pavements. Nevertheless, DWPT continues to be actively researched and developed due to its exclusive benefits: it is not audible, does not harm physical contact, and is universal and can be used with passenger cars, as well as heavy-duty trucks.

This evaluation indicates that TRL is a poor measure of assessing ERS viability. A technology may have a high TRL but be too difficult to use at scale. An example of this limitation can be seen in the Swedish ruling in 2024 to advise against investing in a national ERS network, not due to technical flaws but on the basis of a cost–benefit analysis that made it economically unsustainable unless it were to be widely used by major trading partners, including Germany and France. A more comprehensive index is therefore needed: a Deployment Readiness Level (DRL), which incorporates TRL with the key non-technical considerations, which include economic viability, mass acceptance, regulatory assistance and supply-chain maturity. Catenary systems have a low DRL at the current time, albeit with a high TRL due to their expensive nature and social resistance. On the other hand, inductive systems, which have lower

TRL, can build a higher long-term DRL, overcoming certain deployment barriers. It is rather unlikely that it will be the first technology to achieve technical perfection, which will set the future course of ERS, but rather the first to achieve a high, balanced DRL score in all these dimensions.

Advancing the readiness of all ERS technologies requires addressing several fundamental research gaps:

- **Inductive Efficiency and Power Transfer**: This is most important to reduce the efficiency difference between the conductive system (near 98%) and the inductive system (especially high-power applications: more than 300 kW) needed to support long-haul freight. Even though laboratory demonstrations using more sophisticated technology like parity time symmetric circuits have demonstrated efficiencies as high as 92%, it is a major challenge to transfer this efficiency to the reality of high-speed highway conditions.
- **Interoperability and Standardisation**: Lack of a universal standard of dynamic charging is a key obstacle to the development of transnational networks required to support international freight. As the success of standards like OCPP and ISO 15118 is ready to use in a static way of charging, an integrative model on dynamic conductive and inductive systems is in high demand to avoid market fragmentation and vendor lock-in.
- **Materials Science and Infrastructure Durability**: The concept of charging hardware embedded in the road pavements creates new failure mechanisms such as heat generation, causing thermomechanical stress, interfacial debonding between the charging units and the asphalt, and rapid fatigue of the pavement. More effective initiatives should be conducted to make ERS infrastructure cost-effective and durable in the long term; to achieve this, research on advanced materials like polymer-modified concrete and self-healing asphalt composites is required.
- **Dynamic Energy Control and Management**: The real-time control of power supply to a big fleet of mobile vehicles is a daunting systems-engineering challenge. It requires extremely advanced AI-driven control systems that can handle grid loads, and charge vehicles to the foremost based on their demand and handle billing safely and dependably through millions of micro-transactions.

10.3 Economic and Policy Insights: The Business Case for Electrification

The massive implementation of Electric Road Systems depends on a strong business case and a favourable policy regime. ERS has an economic rationale because it could significantly lower the Total Cost of Ownership (TCO) of commercial vehicle operators, and its implementation requires strategic policy frameworks that would be able to counter the high initial investment costs and fit the national industrial goals.

Economically, the main advantage of ERS is the huge decrease in the size of the on-board batteries, which can take up to 40% of the original price of an electric truck. The factors that allow ERS to allow smaller, less heavy and cheaper batteries include allowing dynamic charging and also doing away with useless downtime in a stationary charging facility. This has a direct impact on enhancing TCO for the fleet operators. The socio-economic cost–benefit studies show that ERS infrastructure is capable of being a lucrative investment for the government, as long as the large initial capital costs are compensated over a sufficiently large volume of vehicle-kilometres, and thus it is optimal to focus on dense freight and public-transport corridors as the most feasible first-use cases. Still, these analyses highlight another issue that becomes a fiscal challenge to governments: the replacement of fossil fuels means that governments lose a substantial amount of revenue in terms of fuel taxes. The possible adverse net present value consideration in terms of finance is what requires the creation of alternative revenue-generation strategies, like road-use fees or electricity fees, to provide the sustainability of the system financially in the long term.

The high initial costs of electrified rail systems (ERS) infrastructure (which are estimated to cost the country billions of dollars to build national-scale networks) create a notable demand to discuss innovative financing models that combine both public and private capital [32]. Public–Private Partnerships (PPP) offer a rather promising variant to overcome this obstacle. BOT or long-term concession arrangements, where a privately owned consortium funds, builds and manages the ERS infrastructure in exchange to receive the right to charge user fees over a specified term, may also be useful when leveraging the efficiency and capital available in the private sector and serve as a key de-risking instrument to offer the initial investment necessary to draw in venture capital.

The strategic case for ERS is further strengthened by its alignment with the industrial and climate policies of major economic blocs:

- **European Union**: ERS implementation is a first-order facilitator of the EU sustainable and smart mobility strategy and the European Green Deal in general. Such policies have ambitious goals, such as ensuring that by 2030, there are at least 30 million zero-emissions vehicles and that there are climate-neutral cities. ERS is a scalable initiative to decarbonise heavy-duty freight on the Trans-European Transport Network (TEN-T), which is one of the priorities to achieve a legally binding set of targets on emissions reduction as stipulated in the EU.
- **India**: ERS has a way out of the extensive emissions of the road-freight sector in India, in line with the objectives of the National Electric Mobility Mission Plan (NEMMP) and the FAME (FASTER Adoption and Manufacturing of Electric Vehicles) scheme [2]. By availing a feasible solution to electrify the trucks and buses, ERS achieves the 2070 net-zero objective of India. Besides, it is aligned with the Make in India program as it opens a prospect of domestic production of the ERS components, such as the power electronics, for advanced paving materials.
- **Japan**: ERS technology is also compatible with the Japanese Green Growth Strategy of selling 100 per cent of new vehicles as electrified by 2035 [43]. Japan has already established itself as a market leader in the automotive segment, with a strong leadership in the field of power electronics and advanced materials. This

country, therefore, stands in a good position to be the major player in the research and development and export of ERS technologies.

10.4 Environmental and Social Impacts: Aligning with Global Sustainability Goals

In addition to its technical and economic superiority, the necessity to establish ERS is deeply based on the possibility of creating significant environmental and social advantages. Causing facilitation of a more efficient, resource-conscious electrification form, ERS is in direct correspondence to global sustainability strategies, such as the Paris Agreement on climate change and the United Nations Sustainable Development Goals (SDGs).

The sustainability case for ERS is built on three key pillars: emissions reduction, resource efficiency, and social equity.

- **Emissions Reduction**: The short-term environmental impact of ERS is the removal of tailpipe emissions of greenhouse gases (GHGs), local air pollutants like nitrogen oxides (NOx) and particulate matter (PM 2.5). This will have a direct effect of alleviating climate change and enhancing human health, especially in urban setups and along major freight trails, as the toxic levels tend to be the greatest in these regions. Full-fledged Life Cycle Assessments (LCAs) affirm that with enhanced renewable electricity grid powering, ERS provides significant GHG reductions over conventional fossil-fuelled vehicles [17], more importantly, ERS improves such advantages by allowing a smaller battery to operate, thereby decreasing the huge carbon and environmental footprint of battery production, which is a significant portion of the overall impact of an EV on the environment.
- **Resource Efficiency**: ERS enables battery downsizing, which is one of its main elements in the resource-efficiency argument. Traditional long-range electric trucks use huge batteries (usually over 600 kWh) to get a decent range that adds weight to the vehicle and decreases the size of the payload, and puts pressure on the availability of lithium, cobalt, and nickel. ERS does not follow this paradigm and lets vehicles work efficiently with significantly smaller batteries suited to the short-range purposes of travelling off the electrified network. This decreased battery addiction addresses the geopolitical risks of the condensed mineral chain of supply and reduces environmental and social deterioration that usually accompany mining activities [18].
- **Social Equity**: ERS can be an effective instrument of promoting social equity, when done in a wise way. Public-transport bus routes can also be electrified to offer cleaner, quieter, and more reliable movement to all groups of people including those with low-income the large portion of the population which frequently relies heavily on the service. Further, as a result of decreasing the number of pollutants caused by increasing heavy-duty trucks, ERS can mitigate the disproportional impact of environmental-health of underserved communities that often reside near

highways, ports, and logistics centers. Nonetheless, to achieve this potential, active policy frameworks will be needed to make such investments in these communities and to make the benefits of cleaner mobility available and affordable to everybody, rather than allowing a flexible division of mobility to appear.

The systemic advantages of ERS overlap with the most important sustainability goals in the world. It offers a practical way through which countries can achieve their Nationally Determined Contribution based on the Paris Agreement through addressing the harder-to-abate road-transport sector that contributes to about 24% of global energy-related CO2 emissions [18]. In addition, ERS is known to contribute to various UN Sustainable Development Goals (SDGs). It directly facilitates SDG 7 (Affordable and Clean Energy) by enhancing the efficiency of energy usage and the incorporation of renewable energy in the transport sector. It helps SDG 9 (Industry, Innovation, and Infrastructure) to grow through the creation of resilient and sustainable infrastructure. It will be vital in meeting SDG 11 (Sustainable Cities and Communities) through the reduction of air and noise pollution within the urban environment and improved clean transport systems that are clean. Lastly, it is an imperative instrument of SDG 13 (Climate Action). This interdependent environment of a sustainable ecosystem is represented in Fig. 10.2.

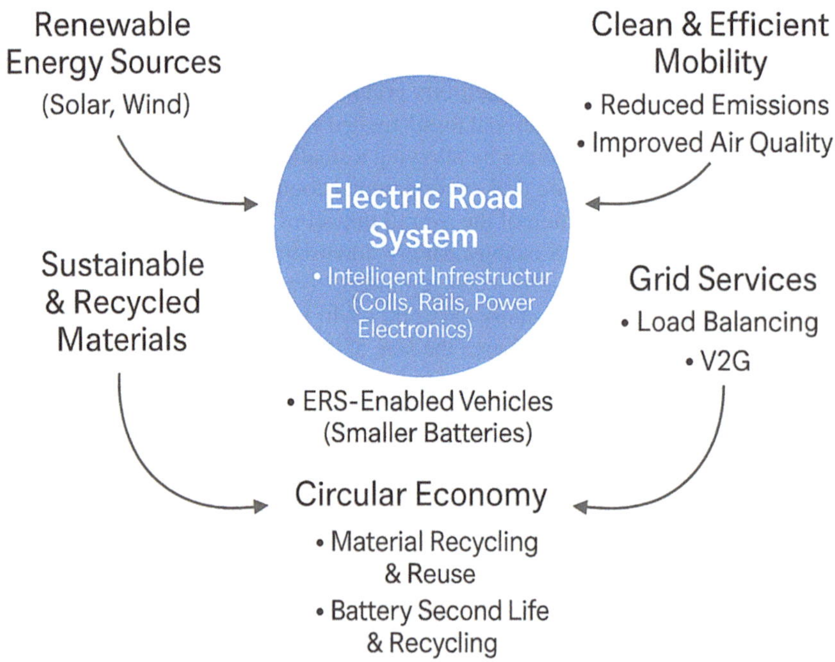

Fig. 10.2 Systems view of a sustainable ERS ecosystem

10.5 Global Strategic Implications: ERS as an Enabler of the Net-Zero Transition

The implementation of the Electric Road Systems is not confined to the national transport policy, but it has immense global strategic consequences on international trade, industrial competitiveness, and technology diplomacy. With countries investing in net-zero-emission pledges, ERS is not only becoming an environmental answer but also a basic infrastructure of a new industrial age. It is strategically important because it can transform the global freight logistics, create new standards regarding technological progress, and change the geopolitical game of the 21st-century clean-energy economy.

ERS is a key accelerator of the net-zero transport transformation, providing one of the limited technologically feasible ways the full decarbonisation of long-haul, heavy-duty road freight can be achieved. Although battery-only electric trucks can be used in shorter routes, they cannot be used in long-haul logistics because the mass and size of required batteries are simply overwhelming, reducing the payload capacity and affecting the charging time. These two drawbacks are bypassed by ERS, which makes it an essential technology for deep decarbonisation in a sector vital to the global economy.

International trade could be radically transformed with the creation of standardised, transnational electrified freight corridors. A cross-continentally distributed network based on interoperable ERS, like the TEN-T network of the EU, may open a chain of seamless, low-cost, and zero-emission logistics and change the competitive landscape of the freight and shipping sectors. Here, technological standardisation is a potent industrial policy/technology diplomacy tool. The countries or trade blocs that effectively establish and export the prevailing ERS standards, be they of a conductive or an inductive system, will have a significant first-mover advantage, therefore dictating the world markets of vehicles, infrastructure components and online management systems. The standardisation process is thus not a technical thing, but a strategic field where future industrial leadership is fought, through such bodies as ISO, IEC and CENELEC.

In the case of developing countries, ERS is a two-sided offer. On the one hand, it provides the possibility to transform the 20th-century paradigm of fossil-reliant transport infrastructure and construct effective, clean, and resilient economic highways starting with the ground. The public transport system and major freight routes could be electrified, which would open a large number of economic and social health benefits. Conversely, ERS infrastructure is characterised by high initial capital, and this barrier is formidable. In the absence of strong technology transfer systems, international climate financing (e.g. multilateral development banks), and technical capacity-building, engineers will risk developing countries falling behind, creating an international mobility chasm between electrified and non-electrified economies.

Finally, the international competition of creating and putting ERS into operation will turn into a new sphere of geopolitical rivalry and strategic partnership. Similar to the creation of 5G telecommunication or semiconductor technology, ERS technology leadership will give one critical power and influence on the design of the supply chains worldwide. Countries that lead in ERS will not just clean up their own transport sector, but will also be in a position to export the high-value technologies, skills, and standards that will form the basis of global mobility in the future.

10.6 Future Research Directions: An Interdisciplinary Agenda for 2025–2045

The transition of Electric Road Systems from pilot projects to a globally deployed infrastructure requires a concerted and interdisciplinary research effort. While foundational concepts have been proven, significant gaps remain across the technological, infrastructural, energetic, digital, and socio-political domains. This section outlines a strategic research agenda for the next two decades, designed to address critical hurdles and unlock the full potential of ERS as a systemic innovation. The agenda is summarised in Table 10.1.

Technology

The core technological challenge is to enhance the performance, efficiency, and cost-effectiveness of power transfer systems, particularly for high-demand applications.

- **High-Speed Dynamic Wireless Power Transfer (DWPT)**: The next major area of research in DWPT is to reach a high efficiency (> 95%) at highway speed (>100 km/h) and high power levels (>300 kW) needed by heavy-duty trucks. This will necessitate the basic research on enhanced coil geometries, new ferrite composite materials with minimal losses at high frequencies, and high-power electronics based on silicon carbide or gallium nitride at high frequencies and robustness.
- **Superconducting Materials**: A high-risk, high-reward, long-term research area is searching into high-temperature superconducting (HTS) coils to be used in WPT. The materials based on HTS might facilitate the near-lossless transfer of power, which will radically change the equation of efficiency. The areas of research should include the creation of more affordable materials in HTS, the creation of compact and efficient cryogenic cooling systems that can be integrated into the vehicle and on the road, and the study of the nature of AC losses of HTS materials under dynamic charges.
- **Hybrid Conductive-Inductive Models**: Studies are needed to create and optimise hybrid models, which exploit the superiority of various technologies. This may include designs in which cars have a very efficient conductive charge at depots or special stations (e.g. buses) and an inductive charge that is flexible during transit on highways. It should be made to focus on seamless switching, on-board power electronics, and system-level cost-performance optimisation.

Table 10.1 Research gaps and priority areas for 2025–2045

Thematic area	Specific research gap	Key research questions	Priority level (2025–2035)
Technology	Low efficiency of high-speed DWPT for heavy-duty vehicles	How can novel metamaterials and coil geometries maintain >95% efficiency at speeds >100 km/h and power levels >300 kW? What are the trade-offs between efficiency, cost, and material durability for new ferrite composites?	High
	Feasibility and cost of superconducting WPT systems	Can compact, vehicle-grade cryogenic systems be developed? What is the true lifecycle cost and energy return on investment for HTS-based ERS?	Low (long-term)
	Lack of integrated hybrid conductive-inductive systems	What is the optimal architecture for on-board power electronics that can seamlessly switch between conductive and inductive charging? How can system-level controllers optimize the use of each mode based on location and grid conditions?	Medium
Infrastructure	Slow and costly on-site construction methods	Can prefabricated, modular ERS pavement sections reduce installation time and cost by over 50%? What automated construction and quality assurance techniques can be developed?	High
	Reactive maintenance models leading to high downtime	What sensor data are most predictive of ERS component failure? Can ML models achieve >90% accuracy in predicting maintenance needs 3–6 months in advance?	High
	Lack of end-of-life strategies for ERS components	What are the most energy-efficient and cost-effective methods for recovering copper, ferrite, and electronics from decommissioned ERS pavements?	Medium
Energy systems	Grid strain from unmanaged high-power charging	What is the optimal mix of dedicated renewables and grid-scale storage for an ERS corridor to achieve > 80% energy self-sufficiency?	High

(continued)

Table 10.1 (continued)

Thematic area	Specific research gap	Key research questions	Priority level (2025–2035)
	Untapped potential of ERS for grid balancing services	What is the aggregate capacity of a 1,000 km ERS corridor to provide frequency regulation? What market mechanisms are needed to compensate vehicle owners for V2G services while in motion?	Medium
	Inefficient static pricing for ERS energy	How can real-time pricing, incorporating grid carbon intensity and traffic data, shift charging loads by >30% away from peak hours?	High
Digital systems	Latency and data overload in centralized control systems	Can a federated AI architecture reduce control loop latency by an order of magnitude compared to a cloud-only model? What is the optimal division of tasks between edge and central AI?	Medium
	Lack of holistic system simulation and optimization tools	Can a comprehensive digital twin model accurately simulate the coupled dynamics of traffic flow, energy demand, and grid response in an ERS corridor?	High
	Security and transparency in multi-operator ERS networks	Can a blockchain-based platform provide a secure and interoperable system for billing and data exchange across international ERS networks?	Medium
Social & policy	Risk of exacerbating mobility inequality	What specific policy interventions (e.g., targeted subsidies, public transit investment) are most effective at ensuring ERS benefits reach low-income communities?	High
	Poor integration of ERS into urban planning	How does the presence of ERS infrastructure influence urban form, land use, and the adoption of shared mobility services?	Medium
	Lack of legal frameworks for transnational ERS corridors	What governance and revenue-sharing models can facilitate private investment in cross-border ERS freight networks?	High

Infrastructure

The durability, cost, and deployment speed of the physical ERS infrastructure are critical for scalability.

- **Modular and Prefabricated Construction**: To fasten the deployment and minimise cost and traffic disruption, a study should be done to create modular ERS infrastructure. This involves prefabricated pavement pieces having inbuilt charging coils or rails, uniformity of utility corridors under the surface of the ground, and automated construction methods. This model can change the ERS implementation process, which is a long civil engineering process, to a more efficient, production-like process.
- **AI-Powered Predictive Maintenance**: Studies are needed to come up with advanced AI and machine learning tools in predictive maintenance of ERS infrastructure. Real-time data on embedded sensors (to check temperature, strain, power output, etc.) and vehicle feedback can be analysed to forecast component degradation and pavement fatigue, allowing proactive maintenance scheduling before failures happen, maximising system uptime and lifecycle cost reduction.
- **Material Circularity and End-of-Life Management**: There is a critical research gap in the ERS end-of-life management. Research is required to come up with cost-efficient and eco-friendly procedures of decommissioning ERS segments, which involve ways of separating and recovering useful materials such as copper, ferrite, and electronic components embedded in asphalts or concrete matrices to recycle into other infrastructure.

Energy Systems

The integration of ERS with the broader energy system must be designed for sustainability and resilience.

- **Deep Renewable Coupling and Energy Storage**: The study should examine the techno-economic viability of setting up energy-positive ERS corridors. This necessitates co-locating ERS infrastructure with focused generation of renewable energy (e.g. solar photovoltaic arrays along highways) and massive battery storage. To reduce dependence on the traditional grid and ensure there is always, 24-h supply of clean energy, modelling of the size and functionality of these integrated systems is necessary.
- **Grid Balancing and Vehicle-to-Grid (V2G) Services**: The prospect of a large group of ERS-linked vehicles to be utilised as an energy asset that is distributed to do grid balancing is great, but much hypothetical. Research should measure the potential with large-scale computation and pilot projects, including the realisation of communication procedures, control algorithms, and market machinery that allow mobile resources to offer ancillary services like frequency regulation, peak demand shaving, and so on.
- **Dynamic Pricing Models**: ERS energy consumption needs advanced dynamic pricing frameworks that are not only based on artificial intelligence, but also dynamic. Such models must go beyond time-of-use rates with real-time grid carbon intensity, local grid congestion, road traffic conditions and vehicle state-of-charge, which will incentivise grid-supportive charging behaviour and efficient demand management.

Digital Systems

The digital layer is the nervous system of an intelligent ERS, enabling its dynamic and optimised operation.

- **Federated AI and Edge Computing**: An autonomous, centralised cloud-based AI system might not be sufficient to handle the mammoth amounts of data and low-latency control demands of a large-scale ERS. Studies ought to focus on federated learning and edge-computing systems or architectures, in which AI models are trained and run at the vehicle or roadside levels, eliminating communication bottlenecks and improving system resilience.
- **Digital Twins**: A high-priority research field is the development of full digital twin models of ERS corridors. These simulated proxies would incorporate real-time information about infrastructure, cars, traffic sensors and the electrical grid, which would be important to include in the early design simulation, optimisation of operation and a what-if situation analysis (e.g., accident response) as well as proactive maintenance scheduling.
- **Blockchain-Based Management and Trading**: The research question that should be addressed is how blockchain technology can be used to establish a safe, open, and decentralised way of running ERS. The possible applications are automated, non-tamperable billing amongst various network operators, secure vehicle-to-infrastructure communications, and allowing peer-to-peer (P2P) energy trading between cars and the power grid.

Social & Policy

Technological and economic feasibility must be matched by social acceptance and robust governance.

- **Inclusive and Equitable Adoption Frameworks**: Policy and Planning Frameworks should be developed that will ensure equitable provision of the ERS benefits. Research needs to be done on how to prioritise the implementation of ERS to minimise pollution in environmental justice communities, generate models of making ERS-enabled public transport more affordable and accessible, and develop plans that would facilitate a just transition among workers in older industries.
- **Urban Mobility Integration and Design**: Case-study research and urban planning modelling are sought in order to understand how to optimally incorporate ERS into the multifaceted structure of the modern city. It goes further than the electrification of current routes and requires the restructuring of the city to cater to multimodal, shared, and sustainable mobility systems based on ERS infrastructure.
- **Transnational Governance Models**: To be followed: To enable ERS to thrive in the international freight business, it is necessary to have strong transnational governance models. Legal and policy studies are needed to establish the frameworks of cross-border investment, technical standards that are harmonised, interoperable billing systems, and revenue-sharing agreements between countries that are involved in international ERS corridors.

10.7 Roadmap for the Future: A Phased Global Rollout

The shift to an electric road system on a mass scale will not happen overnight but will be a gradual process of innovation, standardisation, scaling and integration, taking several decades to be completed. This is a roadmap of a realistic path between 2025 and 2050 that aims to match the major milestones with the international climate and sustainability targets. The development is projected in four stages that overlap each other, as shown in Fig. 10.3.

- **Phase 1: Expanded Pilots & Standardisation (2025–2030)**: This first step will not only go beyond small-scale demonstrators but also roll out multi-technology pilot projects on high-profile freight routes in the major markets in the European Union, the United States and China. The main goals include collecting as much real-life information on performance, durability and economic feasibility and solving the most severe problem of interoperability. The most important accomplishment of this step is the global adoption of a standardised set of requirements of at least one ERS technology (conductive/inductive) so that vehicles produced by other manufacturers and infrastructure can work together. This step coincides with the direct 2030 decarbonisation levels established in the Paris Agreement, which makes ERS a possible technology in the next phase of decarbonisation plans.

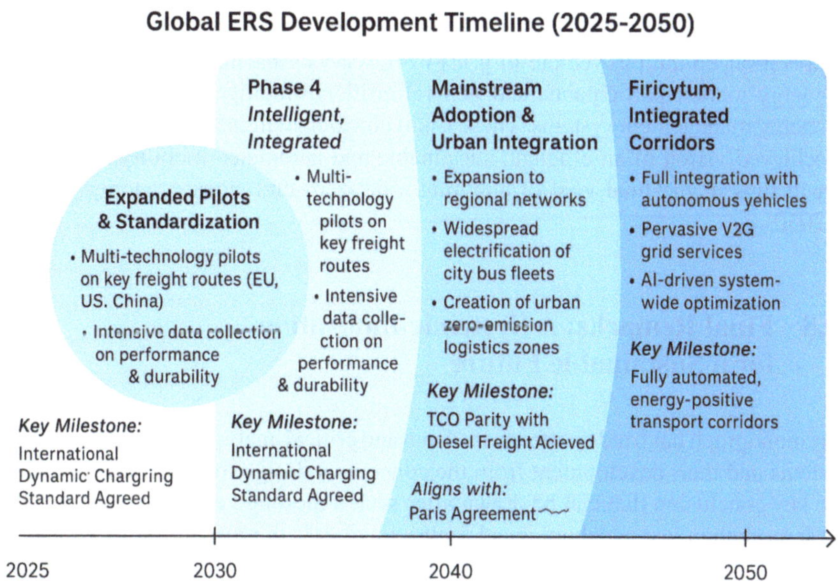

Fig. 10.3 Global ERS (2025–2050)

- **Phase 2: Commercial Scaling on Core Corridors (2030–2040)**: The successful implementation and standardisation of Phase 1 will be followed by the first major commercial implementation of ERS on the most important freight routes in the world during this decade. Technical validation will be supplanted with economic scaling, and the target is to electrify large chunks (e.g., 10–15%) of large systems of networks, including the TEN-T of the EU or the Interstate Highway System of the United States. The milestone will be the attainment of the total cost of ownership equilibrium or better than that of ERS-enabled freight trucks compared to their diesel-powered counterparts, not with the heavy subsidies. This step corresponds to the 2035 and 2040 dates of many countries and regions that intend to exclude the sale of new vehicles with internal combustion engines.
- **Phase 3: Mainstream Adoption & Urban Integration (2040–2050)**: As the technology is proven and the business case is now in place, ERS will go to the mainstream. The use will go beyond major intercity routes to secondary and regional road networks. At the same time, one of the priorities will be the introduction of ERS into the city space, in particular, the electrification of a complete bus park and the development of zero-emission areas in the city logistics. The major milestone of this stage is the standardisation of ERS as the default, domineering answer to novice heavy-duty and civic transport vehicles in the main economies.
- **Phase 4: Intelligent, Integrated Mobility Corridors (2045+)**: In the final phase, the Electric Road System (ERS) will be applied to other disruptive technologies to bring about comprehensively intelligent and integrated transport ecosystems. The infrastructure of ERS will be integrated smoothly with autonomous vehicle operation, ubiquitous vehicle-to-grid (V2G) services that make transport corridors energy market participants, and general artificial-intelligence-based traffic and energy management systems. These smart corridors will incorporate a completely achieved vision of an efficient, sustainable and automated mobility system and will thus fit the final goal of attaining a net-zero emissions economy-wide by 2050.

10.8 Final Remarks: A Systemic Innovation for a Sustainable Future

This monograph has tried to offer an in-depth and critical analysis of the Electric Road Systems and their development from the conceptual design to the emerging reality. The key conclusion that can be drawn after such a thorough analysis is that ERS is much more than just a technological replacement, it is not just an improved way of making an electric car work. Instead, ERS is a systemic innovation, which rearranges the connection between mobility, energy and the built environment fundamentally. It acts as a driver of paradigm shift by forcing the breaking up of silos that have existed throughout the years, dividing transportation planning and energy policy, urban design and digital governance.

The actual value of ERS is in the ability of this system to eliminate the fundamental contradictions of the existing model of transport electrification. It solves the problem of battery addiction, the financial cost of vehicle inactivity and the systemic pressure on the power grids. This way, it will provide a consistent and scalable channel of decarbonisation to the sectors of the economy that have been the most difficult to change, specifically long-haul freight. ERS is converting roads to become active, smart and useful structures.

The way forward, as stipulated in the above road map and research agenda, is admittedly paved with unimaginable technical, financial, and political hurdles. The size of investment is huge, the technological innovations that are necessary are enormous, and the necessity to collaborate internationally and standardise is unquestionable. However, the challenges are not to be regarded as disincentives but the challenge of engineering and policy work of our time. It is a more resilient, efficient and sustainable future that is pursued through ERS. It is the building of the basic infrastructure of the 21st-century clean-energy economy, which has a transformative potential comparable to that of the railways of the 19th or the internet of the 20th. The vision stated in this piece is not a utopian fantasy and is both academically based and technically feasible to make a world where sustainable mobility is achieved by everyone.

References

1. European Parliament (2025) Electric road vehicles in the European Union. https://www.europarl.europa.eu/RegData/etudes/BRIE/2019/637895/EPRS_BRI(2019)637895_EN.pdf. Accessed 18 Oct 2025
2. Invest India (2025) Government's EV policies driving India's green revolution. https://www.investindia.gov.in/team-india-blogs/governments-ev-policies-driving-indias-green-revolution. Accessed 18 Oct 2025
3. Bureau of Energy Efficiency (BEE) (2025) National electric mobility mission plan. https://evyatra.beeindia.gov.in/central-govt-initiative-details/dhi-2/. Accessed 18 Oct 2025
4. U.S. Department of Energy (2025) Federal funding for zero-emission energy and transportation. https://driveelectric.gov/funding-opportunities. Accessed 18 Oct 2025
5. European Commission (2025) Mobility strategy—mobility and transport. https://transport.ec.europa.eu/transport-themes/mobility-strategy_en. Accessed 18 Oct 2025
6. International Energy Agency (IEA) (2025) Global EV outlook 2023—policy developments. https://www.iea.org/reports/global-ev-outlook-2023/policy-developments. Accessed 18 Oct 2025
7. MDPI (2025) Critical review of wireless charging technologies for electric vehicles. https://www.mdpi.com/2032-6653/16/2/65. Accessed 18 Oct 2025
8. Appinventiv (2025) Smart grid for electric vehicle charging infrastructure. https://appinventiv.com/blog/smart-grid-for-elecric-vehicle-charging-infrastructure/. Accessed 18 Oct 2025
9. Bender GmbH (2025) Dynamic load management. https://www.bender.de/en/dynamic-load-management/. Accessed 18 Oct 2025
10. Virta Global (2025) Vehicle-to-grid (V2G): everything you need to know. https://www.virta.global/vehicle-to-grid-v2g. Accessed 18 Oct 2025
11. Wikipedia (2025) Vehicle-to-grid. https://en.wikipedia.org/wiki/Vehicle-to-grid. Accessed 18 Oct 2025

12. GoodFirms (2025) The rise of AI-powered smart roads in the U.S.: are we moving forward? https://www.goodfirms.co/blog/ai-powered-smart-roads-us-traffic-systems. Accessed 18 Oct 2025
13. Google Patents (2025) US10866108B2—systems and methods using artificial intelligence for routing electric vehicles. https://patents.google.com/patent/US10866108B2/en. Accessed 18 Oct 2025
14. AMS India (2025) Eco-friendly roadways: paving the way to a sustainable future. https://amsindia.co.in/eco-friendly-roadways-paving-the-way-to-a-sustainable-future/. Accessed 18 Oct 2025
15. IGI Global (2025) Energy harvesting and smart highways for sustainable transportation infrastructure. https://www.igi-global.com/viewtitle.aspx?TitleId=349758&isxn=9781668492147. Accessed 18 Oct 2025
16. Taylor & Francis Online (2025) Life cycle sustainability assessment of electrified road infrastructure. https://doi.org/10.1080/14680629.2019.1588771. Accessed 18 Oct 2025
17. MDPI (2025) Life cycle assessment of sustainable road networks: current state and future directions. https://www.mdpi.com/2075-5309/13/10/2648. Accessed 18 Oct 2025
18. PubMed Central (PMC) (2025) Life cycle assessment of battery electric vehicles: implications of energy mix and use phase. https://pmc.ncbi.nlm.nih.gov/articles/PMC9171403/. Accessed 18 Oct 2025
19. World Economic Forum (2025) How modular construction drives productivity and circularity. https://www.weforum.org/stories/2025/01/modular-construction-productivity-circularity/. Accessed 18 Oct 2025
20. PIARC (2025) Electric road systems: a solution for the future. https://www.piarc.org/en/order-library/29690-en-Electric%20road%20systems:%20a%20solution%20for%20the%20future. Accessed 18 Oct 2025
21. Taylor & Francis Online (2025) Thermomechanical behaviour of electric road systems: a literature review and experimental findings. https://doi.org/10.1080/14680629.2025.2562222. Accessed 18 Oct 2025
22. DiVA Portal (2025) Maturity of power transfer technologies for electric road systems. https://www.diva-portal.org/smash/get/diva2:1414333/FULLTEXT01.pdf. Accessed 18 Oct 2025
23. Wikipedia (2025) Electric road. https://en.wikipedia.org/wiki/Electric_road. Accessed 18 Oct 2025
24. ResearchGate (2025) Electrification of roads: opportunities and challenges. https://www.researchgate.net/publication/275281180_Electrification_of_roads_Opportunities_and_challenges. Accessed 18 Oct 2025
25. Oxford Academic (2025) Electrification of urban road traffic: a reliability analysis of traction power supply for electric road systems. https://academic.oup.com/tse/article/6/4/tdae003/7596544. Accessed 18 Oct 2025
26. Stanford TechFinder (2025) Efficient dynamic wireless power transfer. https://techfinder.stanford.edu/technology/efficient-dynamic-wireless-power-transfer. Accessed 18 Oct 2025
27. University of Rome Tor Vergata (2025) Challenges and barriers of wireless charging systems. https://art.torvergata.it/retrieve/3d37546a-5a23-4bb9-aa8d-f392ab473586/Palani_Challenges_2023.pdf. Accessed 18 Oct 2025
28. EV CAN (2025) Understanding EV charging interoperability and the standards that support it. https://www.evcan.org/news-insights/understanding-ev-charging-interoperability-and-the-standards-that-support-it/. Accessed 18 Oct 2025
29. Taylor & Francis Online (2025) Large-scale implementation of electric road systems: costs and impact on CO_2 emissions. https://doi.org/10.1080/15568318.2019.1595227. Accessed 18 Oct 2025
30. Delft University of Technology (2025) Prospects for electric road systems on Dutch freight corridors. https://repository.tudelft.nl/file/File_0f2f55f8-fd07-4473-a244-1ec2617a85e0. Accessed 18 Oct 2025
31. IFEU (2025) Opportunities for electric road systems in road freight economics. https://www.ifeu.de/fileadmin/uploads/Publikationen/Mobilit%C3%A4t/CollERS-2-Discussion-paper-3-Opportunities-for-electric-road-systems-in-road-freight-economics.pdf. Accessed 18 Oct 2025

32. DiVA Portal (2025) Socioeconomic analysis of electric road systems. https://www.diva-portal.org/smash/get/diva2:1658492/FULLTEXT01.pdf. Accessed 18 Oct 2025
33. Straits Research (2025) Electric highway (e-Highway) market size, share & trends by 2033. https://straitsresearch.com/report/electric-highway-market. Accessed 18 Oct 2025
34. U.S. Joint Office of Energy & Transportation (2025) Public electric vehicle charging infrastructure playbook. https://driveelectric.gov/ev-infrastructure-playbook. Accessed 18 Oct 2025
35. Sustainability & ESG Professional Blog (2025) Public–private partnership (PPP) models & contract structures. https://oyuntugs.com/en/%F0%9F%A4%9D-public-private-partnership-ppp-models-contract-structures/. Accessed 18 Oct 2025
36. Example Reference (2025) ERS governance PPP models. https://www.example.com/ers-governance-ppp-models. Accessed 1 Jan 1970
37. PubMed Central (PMC) (2025) Economy-wide impacts of road transport electrification in the EU. https://pmc.ncbi.nlm.nih.gov/articles/PMC9380419/. Accessed 18 Oct 2025
38. World Resources Institute (WRI) (2025) India significantly scales up electric vehicle adoption. https://www.wri.org/outcomes/india-significantly-scales-electric-vehicle-adoption. Accessed 18 Oct 2025
39. Changing Transport (2025) Mapping electric vehicle policy for Indian States. https://changing-transport.org/wp-content/uploads/2025_Mapping-electric-vehicle-policy-for-Indian-states.pdf. Accessed 18 Oct 2025
40. NITI Aayog (2025) Status quo analysis of electric mobility and low-carbon passenger transport in India. https://www.niti.gov.in/sites/default/files/2023-03/Status-Quo-Analysis-of-Various-Segments-of-Electric-Mobility-And-Low.pdf. Accessed 18 Oct 2025
41. Agora Verkehrswende (2025) Mapping electric vehicle policy for Indian States. https://www.agora-verkehrswende.org/fileadmin/Projekte/2024/Indian-EV-Policy-Guidelines/115_Indien_EV_Policy_Guidelines.pdf. Accessed 18 Oct 2025
42. Climate Action Tracker (2025) Japan—policies and action. https://climateactiontracker.org/countries/japan/policies-action/. Accessed 18 Oct 2025
43. Bolt.Earth (2025) The rise of electric vehicles in Japan: an exploration of market trends. https://bolt.earth/blog/japan-ev-market. Accessed 18 Oct 2025
44. ResearchGate (2025) Government policy and the development of electric vehicles in Japan. https://www.researchgate.net/publication/222131991_Government_policy_and_the_development_of_electric_vehicles_in_Japan. Accessed 18 Oct 2025
45. Renewable Energy Institute (2025) Assessing the review of electricity system reform: making use of market mechanisms. https://www.renewable-ei.org/en/activities/column/REupdate/20250325.php. Accessed 18 Oct 2025
46. University of Tokyo (2025) Land transportation systems for a sustainable society. https://www.u-tokyo.ac.jp/adm/uci/en/projects/sdgs/projects_00041.html. Accessed 18 Oct 2025
47. Sumitomo Electric (2025) Research & development overview. https://sumitomoelectric.com/sites/default/files/2023-05/download_documents/rd_catalog.pdf. Accessed 18 Oct 2025
48. MDPI (2025) An analysis of electric vehicle charging intentions in Japan. https://www.mdpi.com/2071-1050/16/3/1177. Accessed 18 Oct 2025
49. Journal of Science Policy & Governance (2025) Social equity considerations in the new age of transportation. https://www.sciencepolicyjournal.org/uploads/5/4/3/4/5434385/fleming.pdf. Accessed 18 Oct 2025
50. ResearchGate (2025) Life cycle assessment of electric vehicles—a framework to consider influencing factors. Available at: https://www.researchgate.net/publication/277939015_Life_Cycle_Assessment_of_Electric_Vehicles_-_A_Framework_to_Consider_Influencing_Factors. Accessed 18 Oct 2025
51. Nissan Motor Corporation (2025) Life cycle assessment (LCA). Sustainability. https://www.nissan-global.com/EN/SUSTAINABILITY/ENVIRONMENT/GREENPROGRAM/FOUNDATION/LCA/. Accessed 18 Oct 2025
52. Lund University (2025) Life cycle assessment of electric road systems. https://lup.lub.lu.se/student-papers/record/9088584/file/9089283.pdf. Accessed 18 Oct 2025

53. TRL (2025) ZERFT—ElectroRoad project. https://www.trl.co.uk/projects/zerft---electroroad. Accessed 18 Oct 2025
54. Springer Nature (2025) Nature sustainability—research communities. https://communities.springernature.com/badges/nature-sustainability. Accessed 18 Oct 2025
55. MDPI (2025) Towards the adoption of recommender systems in online sustainable environments. https://www.mdpi.com/2504-2289/9/10/259. Accessed 18 Oct 2025
56. Urban Institute (2025) Promoting equitable development in communities. https://www.urban.org/research/publication/promoting-equitable-development-communities. Accessed 18 Oct 2025
57. Example Reference (2025) Inclusive ERS adoption policy. https://www.example.com/inclusive-ers-adoption-policy. Accessed 1 Jan 1970
58. Union of Concerned Scientists (2025) Low-carbon pathways for transportation. https://www.ucs.org/sites/default/files/2022-09/low-carbon-pathways-for-transportation-white-paper.pdf. Accessed 18 Oct 2025
59. Net Zero Climate (2025) Land transport. https://netzeroclimate.org/sectors/transport/land-transport/. Accessed 18 Oct 2025
60. Nordic Energy Research (2025) Electric road systems. https://www.nordicenergy.org/wp-content/uploads/2019/10/Electric-roads_web.pdf. Accessed 18 Oct 2025
61. International Freight Network (2025) Where global networking leads to success. https://internationalfreightnetwork.com/. Accessed 18 Oct 2025
62. Thomson Reuters (2025) How integrated global trade content transforms compliance. https://tax.thomsonreuters.com/blog/how-integrated-global-trade-content-systems-transform-compliance-into-strategic-advantage/. Accessed 18 Oct 2025
63. U.S. Department of Agriculture (ERS) (2025) Trade policy & world trade organization (WTO). https://www.ers.usda.gov/topics/international-markets-us-trade/trade-policy-world-trade-organization-wto. Accessed 18 Oct 2025
64. VDA (2025) Electromobility and standardization. https://www.vda.de/en/topics/automotive-industry/standardization-and-technical-standards/Electromobility. Accessed 18 Oct 2025
65. University of Coimbra (2025) Standardization in electric vehicles. https://www.uc.pt/en/efs/research/EESEVS/f/XIICLEEE_1844_PereirinhaTrovao.pdf. Accessed 18 Oct 2025
66. Luke Desira (2025) ISO certification for the freight forwarding industry. https://lukedesira.com/industries/freight-forwarders/. Accessed 18 Oct 2025
67. World Bank (2025) Transfer of technology to developing countries. https://openknowledge.worldbank.org/entities/publication/defb7c52-921f-5f2a-90c2-d0de681b8be6. Accessed 18 Oct 2025
68. International Institute for Sustainable Development (IISD) (2025) Technology transfer for climate change adaptation. https://www.iisd.org/publications/report/technology-transfer-climate-change-adaptation. Accessed 18 Oct 2025
69. Stellarix (2025) Wireless power transfer: the future of charging. https://stellarix.com/insights/articles/wireless-power-transfer-the-future-of-charging/. Accessed 18 Oct 2025
70. ResearchGate (2025) Dynamic wireless power transfer in electric vehicles. https://www.researchgate.net/publication/356712790_Dynamic_Wireless_Power_Transfer_in_Electric_Vehicles. Accessed 18 Oct 2025
71. MDPI (2025) A comprehensive techno-economic model for fast and reliable analysis of telecom operator potentials. https://www.mdpi.com/2076-3417/12/20/10658. Accessed 18 Oct 2025
72. PolyU Institutional Research Archive (2025) Overview of superconducting wireless power transfer. https://ira.lib.polyu.edu.hk/handle/10397/110091. Accessed 18 Oct 2025
73. ResearchGate (2025) Analysis of wireless power transfer using superconducting metamaterials. https://www.researchgate.net/publication/329050744_Analysis_of_Wireless_Power_Transfer_Using_Superconducting_Metamaterials. Accessed 18 Oct 2025
74. Hong Kong Polytechnic University (2025) Overview of superconducting wireless power transfer. https://research.polyu.edu.hk/en/publications/overview-of-superconducting-wireless-power-transfer. Accessed 18 Oct 2025

75. ResearchGate (2025) Wireless power transfer using high temperature superconducting pancake coils. https://www.researchgate.net/publication/264568816_Wireless_Power_Transfer_Using_High_Temperature_Superconducting_Pancake_Coils. Accessed 18 Oct 2025
76. Electreon (2025) Our wireless charging electric road projects. https://electreon.com/projects. Accessed 18 Oct 2025
77. Dpublication (2025) Dynamic charging technologies for electric vehicles. https://www.dpublication.com/wp-content/uploads/2022/08/19-20030.pdf. Accessed 18 Oct 2025
78. Vectoral Journal (2025) Comparative study of conductive and inductive charging technologies for electric vehicles. https://vectoral.org/index.php/QJETI/article/download/90/83/178. Accessed 18 Oct 2025
79. ResearchGate (2025) Hybrid topology for domestic conductive and wireless EV chargers. https://www.researchgate.net/publication/391539998_Hybrid_Topology_for_Domestic_Conductive_and_Wireless_EV_Chargers. Accessed 18 Oct 2025
80. ENRX (2025) Wireless inductive charging for buses and trucks. https://www.enrx.com/en/Induction-Applications/Inductive-charging-and-power-applications/Buses-and-trucks. Accessed 18 Oct 2025
81. Example Reference (2025) Conductive and inductive charging duplicate. https://vectoral.org/index.php/QJETI/article/download/90/83/178/. Accessed 1 Jan 1970
82. MDPI (2025) Modular construction in the digital age: a systematic review. https://www.mdpi.com/2075-5309/15/5/765. Accessed 18 Oct 2025
83. MDPI (2025) Special issue: modular railway stations in sustainable transportation systems. https://www.mdpi.com/journal/sustainability/special_issues/U4FKKX88R6. Accessed 18 Oct 2025
84. Dorleco (2025) Predictive maintenance and emobility controls. https://dorleco.com/predictive-maintenance-and-emobility-controls/. Accessed 18 Oct 2025
85. XenonStack (2025) How AI is revolutionizing predictive maintenance for vehicles. https://www.xenonstack.com/blog/ai-predictive-maintenance-vehicles. Accessed 18 Oct 2025
86. Promwad (2025) AI-powered predictive maintenance in electric vehicle charging infrastructure. https://promwad.com/news/ai-predictive-maintenance-ev-charging-infrastructure. Accessed 18 Oct 2025
87. MDPI (2025) Artificial intelligence for predictive maintenance applications: key challenges and future trends. https://www.mdpi.com/2076-3417/14/2/898. Accessed 18 Oct 2025
88. IBM (2025) What is predictive maintenance? https://www.ibm.com/think/topics/predictive-maintenance. Accessed 18 Oct 2025
89. ResearchGate (2025) AI-assisted predictive maintenance of renewable energy infrastructure. https://www.researchgate.net/publication/393414157_AI-Assisted_Predictive_Maintenance_of_Renewable_Energy_Infrastructure. Accessed 18 Oct 2025
90. MDPI (2025) Integration of large-scale renewable energy systems. https://www.mdpi.com/2071-1050/17/3/1311. Accessed 18 Oct 2025
91. Pacific Northwest National Laboratory (PNNL) (2025) Renewable integration. https://www.pnnl.gov/explainer-articles/renewable-integration. Accessed 18 Oct 2025
92. Consensus App (2025) Barriers to integrating renewable energy into the grid. https://consensus.app/search/what-are-the-key-barriers-to-integrating-renewable/XWrx8YtXScyed8MpiZXvAg/. Accessed 18 Oct 2025
93. U.S. Department of Energy (2025) How it works: the role of a balancing authority. https://www.energy.gov/sites/default/files/2023-08/Balancing%20Authority%20Backgrounder_2022-Formatted_041723_508.pdf. Accessed 18 Oct 2025
94. Highland Electric Fleets (2025) What is vehicle-to-grid (V2G) technology? https://highlandfleets.com/vehicle-to-grid-v2g/. Accessed 18 Oct 2025
95. U.S. Department of Energy (2025) Vehicles-to-grid integration assessment report. https://www.energy.gov/sites/default/files/2025-01/Vehicle_Grid_Integration_Asseessment_Report_01162025.pdf. Accessed 18 Oct 2025
96. SICE (2025) Dynamic pricing strategies: a way of improving traffic efficiency and sustainability. https://www.sice.com/en/insigths/dynamic-pricing-strategies-a-way-of-improving-traffic-efficiency-and-sustainability/. Accessed 18 Oct 2025

97. Taylor & Francis Online (2025) Optimal electric vehicle charging with dynamic pricing. https://doi.org/10.1080/03155986.2025.2463189. Accessed 18 Oct 2025
98. ev.energy (2025) Insights into smart charging and dynamic tariffs. https://www.ev.energy/insights. Accessed 18 Oct 2025
99. arXiv (2025) Dynamic pricing and energy management strategy for EV charging stations under uncertainties. https://arxiv.org/abs/1801.02783. Accessed 18 Oct 2025
100. Highways Today (2025) Digital twins are transforming global infrastructure. https://highways.today/2025/09/01/digital-twins-infrastructure/. Accessed 18 Oct 2025
101. Toobler (2025) The role of digital twins in electric vehicles. https://www.toobler.com/blog/digital-twin-for-next-generation-electric-vehicle. Accessed 18 Oct 2025
102. IBM (2025) What is a digital twin? https://www.ibm.com/think/topics/digital-twin. Accessed 18 Oct 2025
103. Autodesk (2025) What is a digital twin? Intelligent data models shaping the built world. https://www.autodesk.com/design-make/articles/what-is-a-digital-twin. Accessed 18 Oct 2025
104. Prism Sustainability Directory (2025) Blockchain applications in community microgrid energy trading. https://prism.sustainability-directory.com/scenario/blockchain-applications-in-community-microgrid-energy-trading/. Accessed 18 Oct 2025
105. ResearchGate (2025) Emergence of blockchain technology application in peer-to-peer electrical energy trading: a review. https://www.researchgate.net/publication/349971963_Emergence_of_blockchain-technology_application_in_peer-to-peer_electrical-energy_trading_A_review. Accessed 18 Oct 2025
106. Example Reference (1970) Blockchain for ERS energy trading. https://www.example.com/blockchain-for-ers-energy-trading. Accessed 1 Jan 1970
107. Cambridge University Press (2025) Design of urban electric bus systems. https://www.cambridge.org/core/journals/design-science/article/design-of-urban-electric-bus-systems/1C0E4AA05F6E1FBF8A545E13F6A8D2DE. Accessed 18 Oct 2025
108. ResearchGate (2025) Analysis and assessment of the electrification of urban road transport based on real-life mobility data. https://www.researchgate.net/publication/325211675_Analysis_and_assessment_of_the_electrification_of_urban_road_transport_based_on_real-life_mobility_data. Accessed 18 Oct 2025
109. MDPI (2025) Electric vehicle integration into road transportation, intelligent transportation, and electric power systems: an Abu Dhabi case study. https://www.mdpi.com/2624-6511/4/3/55. Accessed 18 Oct 2025
110. ResearchGate (2025) Urban mobility in smart cities: a case study in the city of Curitiba. https://www.researchgate.net/publication/353224634_URBAN_MOBILITY_IN_SMART_CITIES_A_CASE_STUDY_IN_THE_CITY_OF_CURITIBA. Accessed 18 Oct 2025
111. Inter-American Development Bank (IADB) (2025) International case studies of smart cities: Rio de Janeiro, Brazil. https://publications.iadb.org/publications/english/document/International-Case-Studies-of-Smart-Cities-Rio-de-Janeiro-Brazil.pdf. Accessed 18 Oct 2025
112. International Energy Agency (IEA) (2025) Global EV outlook 2024—webinar summary. https://iea.blob.core.windows.net/assets/b56dfb05-f547-4912-a66a-007945f0c9da/GEVO2024-Webinar-WEB.pdf. Accessed 18 Oct 2025
113. International Energy Agency (IEA) (2025) Outlook for electric mobility—global EV outlook 2024. https://www.iea.org/reports/global-ev-outlook-2024/outlook-for-electric-mobility. Accessed 18 Oct 2025
114. International Energy Agency (IEA) (2025) Executive summary—global EV outlook 2024. https://www.iea.org/reports/global-ev-outlook-2024/executive-summary. Accessed 18 Oct 2025
115. International Energy Agency (IEA) (2024) Global EV outlook 2024—full report. https://iea.blob.core.windows.net/assets/a9e3544b-0b12-4e15-b407-65
116. ResearchGate (2025) Public charging infrastructure in Japan—a stochastic modelling analysis. https://www.researchgate.net/publication/313831238_Public_charging_infrastructure_in_Japan_-_A_stochastic_modelling_analysis. Accessed 18 Oct 2025

117. Institute for Global Environmental Strategies (IGES) (2025) Mobility and energy policies to achieve net-zero emissions: key opportunities to build a prosperous society. https://www.iges.or.jp/en/publication_documents/pub/nonpeer/en/12605/IATSS+Review+English+Translation.pdf. Accessed 18 Oct 2025
118. Joint Office of Energy and Transportation (2025) National electric vehicle infrastructure formula program annual report: plan year 2023–2024. https://driveelectric.gov/files/nevi-annual-report-2023-2024.pdf. Accessed 18 Oct 2025
119. Advanced Energy United (2025) Key 2024 legislative trends steering the future of electrified mobility. https://blog.advancedenergyunited.org/2024-ev-state-policy-wrapped. Accessed 18 Oct 2025
120. Council of European Energy Regulators (CEER) (2025) Electric vehicles: network management and consumer protection. https://www.ceer.eu/wp-content/uploads/2023/08/EV-Report.pdf. Accessed 18 Oct 2025
121. Japan Science and Technology Agency (JST) (2025) Agency overview. https://www.jst.go.jp/EN/about/pdf/outline_e.pdf. Accessed 18 Oct 2025

MIX
Papier aus verantwortungsvollen Quellen
Paper from responsible sources
FSC® C105338

If you have any concerns about our products,
you can contact us on
ProductSafety@springernature.com

In case Publisher is established outside the EU,
the EU authorized representative is:
Springer Nature Customer Service Center GmbH
Europaplatz 3, 69115 Heidelberg, Germany

Printed by Libri Plureos GmbH
in Hamburg, Germany